AF352689

Teaching the Literature of Climate Change

Teaching
the Literature
of Climate Change

Edited by

Debra J. Rosenthal

The Modern Language Association of America
New York   2024

To order MLA publications, visit www.mla.org/books. For wholesale and international orders, see www.mla.org/bookstore-orders.

The MLA office is located on the island known as Mannahatta (Manhattan) in Lenapehoking, the homeland of the Lenape people. The MLA pays respect to the original stewards of this land and to the diverse and vibrant Native communities that continue to thrive in New York City.

Options for Teaching 64
ISSN 1079-2562

Library of Congress Cataloging-in-Publication Data

Names: Rosenthal, Debra J., 1964– editor.
Title: Teaching the literature of climate change / edited by Debra J. Rosenthal.
Description: New York : The Modern Language Association of America, 2024. | Series: Options for teaching, 1079-2562 ; 64 | Includes bibliographical references.
Identifiers: LCCN 2023050168 (print) | LCCN 2023050169 (ebook) | ISBN 9781603296342 (hardcover) | ISBN 9781603296359 (paperback) | ISBN 9781603296366 (EPUB)
Subjects: LCSH: Literature—Study and teaching (Higher) | Climatic changes—Study and teaching (Higher) | Environmental literature—Study and teaching (Higher) | Ecofiction—Study and teaching (Higher) | Climatic changes in literature. | BISAC: LANGUAGE ARTS & DISCIPLINES / Writing / Authorship | SCIENCE / Global Warming & Climate Change | LCGFT: Literary criticism. | Essays.
Classification: LCC PN59 .T446 2024 (print) | LCC PN59 (ebook) | DDC 363.70071/1—dc23/eng/20231229
LC record available at https://lccn.loc.gov/2023050168
LC ebook record available at https://lccn.loc.gov/2023050169

Contents

Part III: Texts

Part IV: Courses and Interdisciplinarity

Part V: Assignments

Part VI: Hopefulness and Beyond

Debra J. Rosenthal

Introduction

Now that we are two decades into the twenty-first century, courses that thematize literature's engagement with issues of anthropogenic climate change have become increasingly popular and, surely, more of an ethical imperative to teach. As high school, college, and graduate students realize that their generation's future will be drastically affected by climate disruption, creative artistry becomes ever more relevant in helping readers grasp the existential threat to place, identity, and culture. Many educational institutions offer courses on climate fiction, or cli-fi, and environmental literature that are taught by instructors whose literary research and personal concerns align with their pedagogical interests. Although many books address strategies for teaching environmental history or climate change from an interdisciplinary perspective, we need more scholarship devoted to the teaching of fiction, poetry, and drama that deeply engages our carbon economy and global heating.

Students want and need to understand the global environmental devastation they will inherit. The literary imagination powerfully addresses such consequences as warming temperatures, ocean acidification, desertification, sea-level rise, climate refugees, the spread of disease, feedback loops, and the collapse of our biome and thus has become an essential tool for constructing meaning in an age of climate change. Many poets, novelists, and playwrights engage what the peace-building expert John Paul Lederach calls "the moral imagination"—courage and discernment to bring about cultural transformation that values humanity over profits and utility. This moral imagination, this serious literary engagement with the threat of global disaster, is crucial to convey to students as they develop their own critical thinking skills and ethical compass.

Instructors can easily fill syllabi with creative works that address what Sylvia Mayer calls "narratives of anticipation"—showing our world on the verge of climate collapse—and "narratives of catastrophe"—depicting a world where climate collapse has already occurred (13). But instructors in the literature classroom may struggle to figure out course design, remote instruction, pedagogical strategies, texts that speak to their students and to one another, strategies for approaching such materials, and the sorts of assignments that might generate the most meaning. The contributors to this volume have thought deeply about these issues and in these pages share broad-ranging ideas from their extensive experience teaching the literature of climate change.

As literary scholars, the contributors also publish their research widely and present at scholarly conferences. Important journals in the field, such as *Interdisciplinary Studies in Literature and Environment, Resilience: A Journal of the Environmental Humanities,* and *Ecozon@: European Journal of Literature, Culture, and Environment* regularly publish many essays on the literature of climate change. Conferences around the world feature panels devoted to the literary imagination's representation of climate distress. The contributors' essays jibe well with other relevant texts in the field, such as Amitav Ghosh's *The Nutmeg's Curse,* Roy Scranton's *We're Doomed. Now What? Essays on War and Climate Change* and *Learning to Die in the Anthropocene: Reflections on the End of a Civilization,* Kathleen Dean Moore's *Great Tide Rising: Toward Clarity and Moral Courage in a Time of Planetary Change,* Stephen Siperstein's *Teaching Climate Change in the Humanities* (Siperstein contributes a chapter to this volume), Bill McKibben's *The Global Warming Reader: A Century of Writing about Climate Change,* David Wallace-Wells's *The Uninhabitable Earth,* and many others.

But the work that goes into conducting literary research is very different from the work that goes into preparing to teach literature in the classroom. Since literary critics and scholars want to make their areas of expertise accessible to students, this volume serves as a forum where instructors can learn from one another about best pedagogical strategies for both in-person and remote teaching as well as a resource for instructors eager to develop such courses for their students. Some of the essays discuss big-picture aspects of teaching, and others go into more detail about specific class activities, including remote-teaching options.

The discipline of literary studies can offer a space for thinking about the climate crisis that is distinct from other fields of inquiry. In that space, instructors can think beyond the norms of traditional response papers or literary essays. For example, the conceptual challenges and emotions elicited by climate change might be best suited to creative assignments such as personal essays, poetry, or fiction. Creating and studying such works illuminates the power of literature to create consciousness and community around shared predicaments. For example, as the COVID-19 crisis highlights the connections between a global pandemic and climate change, it prompts consideration of the role of human action and inaction in instigating, exacerbating, and potentially ameliorating the twin crises. The literature classroom is poised to ask students what kinds of narratological elements writers use to convey urgency in this situation and how writers imagine and construct their readers as participants in a crisis. The literature classroom can also uniquely contribute to the study of climate change by helping students understand how voices of empathy and sympathy can represent and extend care and resources. On the flip side, literature instructors can encourage awareness of the slippery tendency of empathy-building to veer into sentimentalizing suffering, which can distance the reader from people's real lived experiences of harm. Cli-fi shows the power of the literary imagination to envision a way out of or beyond the problem, but instructors need to have a handle on what, pedagogically, makes the difference between students creatively engaging with solutions and using disaster literature as lurid escapism.

Indeed, pursuing equitable solutions to the climate crisis demands the understanding of rhetoric and audience that literary studies can supply. For example, instructors can help students see the deep history of the metaphors through which we talk about climate impacts. Some terms associated with sea-level rise, for instance, such as "drowning" and "flooding," were used in early theories of eugenics, especially in Roman society, which enfranchised patriarchs to drown children born with disabilities. American eugenics often discussed the fear of being "drowned" by a "flood" of lower intelligence as a justification for sterilization. Those who believe in white superiority often refer to a "flood" of nonwhite immigrants in calls to tighten border security regulations. These terms uncomfortably overlap with sea-level rise and the flooding that will predominantly occur in communities of color, coastal communities, and low-lying nations or island nations (often where fossil fuels are extracted or refined). Such rising sea levels will cause massive climate refugee migration that will indeed "flood" countries that have more resources. Language and literature instructors are poised to discuss the rhetoric of climate racism, in which Black and brown people in low-lying areas are seen to have less value, and so their homes and livelihoods are seen as expendable.

Ironically, sometimes our aim of teaching about climate change on a global scale can conflict with other pedagogical values. For example, institutions that want their students to become interculturally aware promote numerous study abroad programs. Educators tend to agree with Janie in Zora Neale Hurston's *Their Eyes Were Watching God*: "[Y]uh got tuh go there tuh know there" (285); since we want our students to comprehend the dazzling array of human knowledge and customs, we send them abroad to experience other cultures. However, the act of creating opportunities dependent on international airplane travel increases students' carbon footprint and thus stands in stark contradiction to our simultaneous messages about the very planetary stewardship we urgently want to inculcate. This problem is particularly egregious since, according to Richard Slimbach, a professor of global learning at Azusa Pacific University, over sixty percent of study abroad programs are of weeklong duration over breaks rather than providing a full semester or year of cultural immersion. Slimbach questions the ethics of global travel in an age of climate catastrophe: "Especially in programs of compressed duration, it is hard to see how the educational benefits might adequately compensate for the environmental harms" (qtd. in Redden). Similarly, the shift to online teaching during the COVID-19 pandemic illuminated for climate-oriented instructors that remote learning can reduce the use of fossil fuels required for students to commute to class. Planetary climate catastrophe demands that we think hard about how to educate students to be global citizens. This volume of essays aims to help colleagues think of best practices when setting up courses to teach about our carbon economy.

Part 1 of this volume, "Principles," addresses key concepts and strategies that can help instructors think about ways to address climate change in literature. Stef Craps shows how questions of environmental justice continually guide and inform classroom discussions at the graduate level, shedding light not only on texts that explicitly engage with such concerns but also on texts that largely evade them. Andrew Hageman articulates the opportunities and limitations of exploring "weird fiction" in climate change literature courses. April Anson explores strategies for teaching the literature of climate change centered on the connection of social and environmental justice. Ted Martinez positions climate change literature as a gateway to climate change learning and action, demonstrating that engagement with literature can change readers' perceptions of climate problems and solutions. Stephen Siperstein demonstrates the potential of nonliterary cultural forms, such as games, to foster creativity and community in a cli-fi course. Jo Alyson Parker explains why she teaches climate change literature as a first-year seminar and how she adapted her curriculum to engage students' experience of a global pandemic. Observing that many recent stories aimed at young adults are set against a backdrop of climate change, Sofia Ahlberg recommends ways to teach cli-fi in the coming-of-age genre. In his course, Matt Burkhart uses Rob Nixon's ideas about "apprehension" to help students identify the beliefs, values, and practices that have led to our current juncture as well as narrative techniques that can help arrest those trajectories and initiate paths that generate meaning.

Part 2, "Locations," discusses courses that focus on particular bioregions to help instructors interested in teaching how our carbon economy will impact different ecozones. The first three essays address island nation literature. Christina Gerhardt takes readers through various texts she uses to consider Caribbean literature, Black feminist writing, and climate change. Clare Echterling explains her online pedagogy for teaching anti-colonial literature about climate change and focuses on works by the Marshallese poet, performance artist, and activist educator Kathy Jetñil-Kijiner. Mary Ann Gosser-Esquilín presents strategies for teaching the first Caribbean econovel, Mayra Montero's *In the Palm of Darkness*. Nassim W. Balestrini turns to the Arctic, teaching Chantal Bilodeau's *The Arctic Cycle* in ways that challenge students to explore questions about Western versus Inuit conceptualizations of space and sociality and their respective impact on geographically demarcated ecosystems. In Finland, Parker Krieg faces the challenge of bridging disciplinary interests and national cultures by approaching climate change literature through a consideration of the relationship between professionals and publics. Thomas Hallock brings the discussion back to the United States as he describes how he has shifted the focus of his nature writing classes to study the fragmented creeks and culverts that lie close to campus.

Part 3, "Texts," offers strategies for specific works that instructors have found most engaging. Magdalena Mączyńska uses the dialogism of Barbara Kingsolver's *Flight Behavior* to allow students to develop a meta-awareness of challenges involved in discussing climate change across disciplinary and ideological fields; by juxtaposing multiple subject positions in various degrees of tension, Kingsolver's text invites readers to reconsider their own positionality through the critical lens of another. Teresa Goddu focuses on three key problems in the teaching of climate fiction: how to make the case for literature's role in the climate conversation, how to organize a course around a rapidly emerging and still developing genre, and how to enact pedagogical strategies that cultivate the capacities students will need to face the climate crisis. Robert Marzec lays out a strategy for teaching texts from Margaret Atwood's *MaddAddam* trilogy through closer study of the impact of cultural forces on our ability to confront the crisis of climate change and to interact with nonhuman intelligence. Jason Molesky discusses teaching creative climate nonfiction ("cli-nofi") in a classroom of incarcerated college students. To close this section, Aaron Rosenberg uses Virginia Woolf's *Orlando* to help students think about the link between global warming and colonial violence.

In part 4, "Courses and Interdisciplinarity," the volume then considers interdisciplinary and team-teaching approaches to engaging with the literature of climate change. Hannah Kroonblawd's essay describes how, through reading and writing in a variety of genres, her students consider questions of worldview, focalization, sociocultural influences, history, and local, national, and transnational contexts. Patrick Whitmarsh describes a course that helps students think critically about global warming, question the normative histories of industrial and cultural progress, and contemplate uncertainty not as a cause for paralysis but as a catalyst for change. Cynthia Williams's environmental humanities course offers a valuable model for teaching the literature of climate change in a polytechnic setting—where a very pragmatic student body is preparing for the hands-on work of adaptation and mitigation. Ali Brox provides strategies and practical examples from a first-year cli-fi seminar and an introductory environmental studies course that integrates the literature of climate change into a unit or module that complements other environmental topics and issues. This section of the volume concludes with two essays on courses taught in tandem. Debra J. Rosenthal and Jeffrey Johansen describe teaching a cli-fi course whose students must co-enroll in a biology class on climate change. In a summative linked project that draws upon both biology and literary studies, students must make a pitch for a Hollywood-style movie blockbuster about climate change. At another university, Ben Jamieson Stanley teaches Climate Fiction and Environmental

Refugees in conjunction with Emily S. Davis's course Moving Fictions; the former class focuses on environmental justice, the latter emphasizes migration, and the co-enrolled students build pages on environmental refugeeism for a *Moving Fictions* website.

Part 5, "Assignments," presents assignments that help open up texts and topics for students. In the first essay, Orchid Tierney taps into the power of podcasts to help students foreground local knowledge that might otherwise be erased in more traditional forms of research publication. Melissa Anderson shares an assignment for teaching the critical importance of information literacy and the discourse of science through comparison of contemporary news sources and works of fiction in the ways they establish authority for scientific content. Barbara Leckie elaborates on three pedagogical approaches—introductions focused on "expertise," "keyword" work groups, and a one-day "student-selected syllabus"—aimed at teaching the forms that climate change takes in a range of cultural works and encouraging a form of response adequate to the complexity and urgency of the topic. Finally, Tobias Menely's assignment integrates Representative Concentration Pathways (RCPs) with literary study, asking students what sort of trajectory a given novel might be following as they consider the RCPs as themselves a form of narrative presentation.

The final section, "Hopefulness and Beyond," moves toward a position of hope through five essays and an afterword that explore students' responses to climate change and ways to develop an encouraging mindset in the face of bleakness. As fires raged not far from where she taught in Australia, Kathryn Prince's cli-fi class started to look less like science fiction and more like science fact; in response, she uses her literature class to "inoculate" students against climate despair and help them think through ways to engage and respond productively. In her class, Brianna Burke counters students' emotional or intellectual fatigue through writing assignments, opportunities for emotional expression, and mini-segments that highlight the miraculous nature of our planet and hold up examples of people and organizations working for cultural, environmental, or planetary change. Ria Banerjee discusses her teaching experience of transforming the ubiquitous question "So what?" into "Now what?" as students consider technological advances like posthumanism that might hold the potential to counter the worst effects of climate change and they respond in creative and critical ways to define "the human of tomorrow." Rick Van Noy describes creative projects that can lead students out of the despair that can arise from depressing texts. Recognizing student hopelessness in the age of climate crisis, Jennifer Atkinson outlines a collective storytelling assignment that helps students generate visions of alternative futures. Finally, in an afterword, Sarah Jaquette Ray reflects on the pedagogical work of the volume as a whole and meditates on the need to slow down long enough to notice the urgency of climate devastation and identify the best courses of action.

Works Cited

Ghosh, Amitav. *The Nutmeg's Curse: Parables for a Planet in Crisis.* U of Chicago P, 2021.

Hurston, Zora Neale. *Their Eyes Were Watching God.* U of Illinois P, 1978.

Lederach, John Paul. *The Moral Imagination: The Art and Soul of Building Peace.* Oxford UP, 2010.

Mayer, Sylvia. "Explorations of the Controversially Real: Risk, the Climate Change Novel, and the Narrative of Anticipation." *The Anticipation of Catastrophe: Environmental Risk in North American Literature and Culture,* edited by Mayer and Alexa Weik von Mossner, Universitätsverlag Winter, 2014.

McKibben, Bill. *The Global Warming Reader: A Century of Writing about Climate Change.* Penguin Books, 2012.

Moore, Kathleen Dean. *Great Tide Rising: Toward Clarity and Moral Courage in a Time of Planetary Change.* Counterpoint, 2016.

Redden, Elizabeth. "International Education in an Era of Climate Change." *Inside Higher Ed,* 19 Dec. 2019, www.insidehighered.com/news/2019/12/19/international-educators-begin-confront-climate-crisis.

Scranton, Roy. *Learning to Die in the Anthropocene: Reflections on the End of a Civilization.* City Lights Publishers, 2015.

———. *We're Doomed. Now What? Essays on War and Climate Change.* Soho Press, 2018.

Siperstein, Stephen. *Teaching Climate Change in the Humanities.* Routledge, 2016.

Wallace-Wells, David. *The Uninhabitable Earth: Life after Warming.* Tim Duggan Books, 2019.

Part I

Principles

Stef Craps

Climate Justice and
the Literary Imagination

Questions of climate justice tend not to play a prominent role in international political discussions such as UN climate summits, which are usually dominated by economic and technical measures for mitigation and adaptation. The adjective *global* in the term *global warming* implies that the entire world is affected by climate change, and scientific reports such as the ones periodically released by the Intergovernmental Panel on Climate Change focus on average increases in global temperatures. We are all in the same boat, it seems.

But in truth, of course, we are not: there are major inequalities in the global distribution of responsibility for and vulnerability to climate change. Those least responsible for climate change tend to be hit the hardest by its impacts. The West, which is responsible, historically, for most greenhouse gas emissions, is least vulnerable; the Global South is most vulnerable. In addition to geographical location, vulnerability to climate change is determined by factors such as race, gender, and socioeconomic status: people of color, women, and poor communities are more likely to be affected than white people, men, and rich communities.

Insofar as it buys into the Anthropocene narrative, climate change scholarship in the humanities and social sciences risks participating in the tendency to depict global warming as if it affected all humans equally, regardless of race, gender, class, or national differences. The Anthropocene is a hypothesis advanced in 2000 by the chemist Paul Crutzen and the biologist Eugene Stoermer, who argued that the Holocene—the postglacial epoch that began approximately 11,700 years ago—was over and that the earth had entered a new geological epoch driven by human activity. A common criticism of this popular idea is that it obscures questions of climate justice. As the postcolonial ecocritic Rob Nixon puts it, "[T]he Anthropocene's grand species perspective on the human . . . risk[s] suppressing—historically and in the present—unequal human impacts, unequal human agency, and unequal human vulnerabilities" ("Anthropocene"). Nixon speaks of the need to counter "the centripetal force" of the dominant story of the Anthropocene as a grand species narrative with "centrifugal stories" that acknowledge these immense disparities.

That is why some critics of the Anthropocene have coined alternative terms such as *Capitalocene* (Malm; Moore) and *Plantationocene* (Haraway et al.; Tsing). They blame the climate crisis and the broader ecological crisis not on an abstract humanity but on the capitalist mode of production or the plantation system and the specific societies adopting it. The Indigenous scholar Kyle Whyte, citing Heather Davis and Zoe Todd, declares that "the Anthropocene is rooted in colonization" and that climate change is basically "an intensification of colonialism" ("Indigenous Climate Change Studies" 159, 156). Discussions of climate change should be informed, then, by a postcolonial sensibility: we need to "decolonize" the Anthropocene, in the parlance of the day (also used by Whyte, as well as by Davis and Todd), lest we forget the historical processes that got us into this mess and that account for the uneven distribution of climate change impacts and the unequal conditions of life in the current era.

This view is articulated very clearly and eloquently by Naomi Klein, one of the most prominent voices in the climate movement, in an essay titled "Let Them Drown: The Violence of Othering in a Warming World." She argues that insights from postcolonial theory, and particularly the work of Edward Said, are highly relevant for understanding climate change and can help us respond to it. Klein finds Said's concept of othering, which he defined as "disregarding, essentialising, denuding the humanity of another culture, people or geographical region," especially inspiring. She contends that othering is inherent to the kinds of exploitation of resources and people that have led to the climate crisis:

> [T]he thing about fossil fuels is that they are so inherently dirty and toxic that they require sacrificial people and places: people whose lungs and bodies can be sacrificed to work in the coal mines, people whose lands and water can be sacrificed to open-pit mining and oil spills. . . . There must be theories of othering to justify sacrificing an entire geography—theories about the people who lived there being so poor and backward that their lives and culture don't deserve protection.

Among the theories of othering that Klein mentions are Manifest Destiny, Terra Nullius, and Orientalism. Adamant that the climate crisis is not a crisis of human nature, Klein is wary of the notion of the Anthropocene insofar as it seems to suggest just that and lets systems such as capitalism, colonialism, and patriarchy off the hook. Klein points out that the process of othering also facilitates the waging of wars for oil in regions like the Middle East, since othering nations and peoples effectively deprives them of the right to control their own oil in their own interests. Moreover, according to Klein, othering explains why Western countries have no qualms about carrying out drone strikes in conflict zones along the aridity line in the Middle East and North Africa. And if the inhabitants of these places become refugees, they are dehumanized yet again: "their need for security" is cast as "a threat to ours" and "their desperate flight" as "some sort of invading army."

Once we understand the connections between the various systems of othering that sustain our present-day reality, we can tackle the climate crisis more efficaciously. Klein argues that, while climate change acts as "an accelerant to many of our social ills," it can also be "an accelerant for the opposite"—that is, for the forces working for justice: "[T]he climate crisis . . . might just be the catalyst we need to knit together a great many powerful movements, bound together by a belief in the inherent worth and value of all people and united by a rejection of the sacrifice zone mentality, whether it applies to peoples or places." Climate change is not only a pressing problem but also an opportunity to create a more just and sustainable world.

Two scholars who have recently looked at literary engagements with issues of climate justice, Matthew Schneider-Mayerson and Antonia Mehnert, share the view that literature has a vital contribution to make to the climate justice conversation because of its ability to generate empathy for "people across time, and thus future generations, as well as with people in different social, economic, and ethnic contexts" in the present (Mehnert, qtd. in Schneider-Mayerson 948). Literature can expose and challenge Klein's sacrifice zone mentality by making visible and promoting empathy with the different perspectives of silenced others. Schneider-Mayerson writes that "climate fiction can play a powerful role in influencing the frames that readers perceive, prioritize, adopt, and share with family, friends, coworkers, and others. The novel in particular has great potential to encourage and cultivate transnational empathy for the already-disadvantaged victims of climate change" (961). Mehnert concurs that "[c]ultural productions such as films or literature can . . . serve as key sites that contest universalizing GHG narratives because they provide 'insider perspectives' on the struggle for climate justice and reveal what otherwise remains hidden in emission graphs—that is, the intra-national, social, and ultimately personal dimensions of environmental injustice" (189).

It is striking, though, that as yet there appear to be very few studies of climate change literature that actually focus on climate justice or even just pay attention to it. Schneider-Mayerson notes that "surveys of the genre or category of 'cli-fi' rarely include a mention of climate justice" (962 n2). In his book *Slow Violence and the Environmentalism of the Poor*, Nixon observes that the relationship between environmentalism and postcolonialism "has been, until very recently, dominated by reciprocal indifference or mistrust," despite the fact that both fields—which emerged around the same time and are among the most dynamic in literary studies—"have both exhibited an often-activist dimension that connects their priorities to movements for social change" (233). He attributes the lack of interaction to "four main schisms . . . between the dominant concerns of postcolonialists and ecocritics": an attachment to hybridity and cross-culturation versus purity, a concern with displacement versus place, a preference for the cosmopolitan and transnational versus the national, and an interest in history versus timeless transcendentalism (236). While ecocriticism, in particular, has gradually moved away from these early priorities in recent decades, and postcolonial ecocriticism is now one of its most thriving subfields, the aforementioned lack of attention to issues of climate justice in scholarship on climate change literature could be a lingering symptom of the historically fraught relationship between the two fields.

Schneider-Mayerson proposes an alternative explanation in the subtitle of his article "Whose Odds? The Absence of Climate Justice in American Climate Fiction Novels." Indeed, he claims that climate justice concerns are all but absent in recent American climate fiction and that that is why they hardly feature in literary-critical responses to it: there is nothing much to report to begin with. He acknowledges that writers like Octavia E. Butler and Paolo Bacigalupi have produced literary texts that focus on climate justice—some of which Mehnert discusses under that rubric in her book *Climate Change Fictions: Representations of Global Warming in American Literature*. However, he argues that these works constitute only "a small minority" of recent American climate fiction. According to Schneider-Mayerson, in the early twenty-first century, "[W]ell-known authors generally shied away from depicting in detail the violence (both slow and spectacular) of climate injustice" (958). They chose to depict climate change "in specific, limited, and surprisingly problematic ways," portraying it as "a problem for white, wealthy, educated Americans" and "secondarily gestur[ing] toward its consequences for human beings in general—the monolithic and flattened 'we' of *homo sapiens*" (945). Thus, they effectively "ignored climate justice, . . . reflect[ing] and potentially reif[ying] a narcissistic tendency among many white American readers" (945).

Schneider-Mayerson makes his case through a close reading—which takes up most of the article—of what he considers to be "two representative texts that have been widely reviewed, assigned, and analyzed and are therefore likely to have reached a large number of readers" (945): Nathaniel Rich's *Odds against Tomorrow* and Kim Stanley Robinson's *Science in the Capitol* trilogy. He is convinced that, small though this sample may be, "we would come to a similar conclusion by examining the vast majority of climate fiction published during this period" (958).

I agree with Schneider-Mayerson that it is "critical to keep justice firmly in mind" and that "readers, critics, publishers, scholars, and teachers" therefore "ought to ask of every climate change narrative, in literature and other media: whither climate justice?" (961). In fact, that is exactly what I try to do myself when teaching climate change fiction, which I have been doing for the last several years. At Ghent University, I teach a graduate course on the literary imagination of the climate crisis in which we concern ourselves with issues of climate justice as they manifest themselves (or hide themselves, as the case may be) in the literary and other artistic works under discussion. One of the guiding questions throughout the course is whether race, gender, socioeconomic status, and geographical location factor into these works' engagement with climate change or are obscured. I try to cultivate a sensitivity to questions of climate justice and, when such questions are not thematized, which is indeed often the case, to highlight and critically interrogate the apparent absence of such a concern.

For example, in a class on humorous climate change literature that focuses on Ian McEwan's novel *Solar*, we consider the implications of the fact that a wealthy, white Englishman who incarnates everything that is reprehensible about modern man acts as an allegorical Everyman figure, through whom the novel suggests that "human nature" is responsible for the climate crisis and our inaction in the face of it. Moreover, we reflect on the question of whether climate change humor is enabled by privilege, and, if so, whether this should give us pause. After all, comedic treatments of climate change—besides *Solar*, we also discuss the title story from Helen Simpson's short story collection *In-Flight Entertainment* (3–23)—tend to be focalized through characters who are privileged in every way; there would appear to be less humor to be found in narratives from the perspectives of poor, nonwhite, non-Western, or female characters for whom climate change is no laughing matter, as they bear the brunt of it.

In another class, we discuss climate anxiety in relation to Jeff Nichols's film *Take Shelter* and—albeit less extensively—Paul Schrader's film *First Reformed*. Both films feature leading characters who are overwhelmed with dread about climate change; in fact, they appear to be suffering from pretraumatic stress disorder: they are traumatized, it seems, by imagining future climate catastrophe. In this case, too, I make a point of highlighting the fact that the victims of pretraumatic stress disorder in these fictional examples are all white, male Americans, which raises the question of how race, gender, and geopolitical location factor into this diagnosis. Pretraumatic stress disorder would appear to be associated with a position of privilege that has thus far provided physical protection from the disastrous consequences of climate change that are already being experienced by many less fortunate people around the world. Indeed, as Whyte argues, "Climate injustice, for Indigenous peoples, is less about the spectre of a new future and more like the experience of déjà vu" ("Is It Colonial Déjà Vu?" 88). Davis and Todd describe a "seismic shockwave of colonial earth-rending" that

> has rolled through and across space and time and is now hitting those nations, legal systems, and structures that brought about the rending and disruption of lifeways and life-worlds in the first place. The Anthropocene—or at least all of the anxiety produced around these realities for those in Euro-Western contexts—is really the arrival of the reverberations of that seismic shockwave into the nations who introduced colonial, capitalist processes across the globe in the last half-millennium in the first place. (774)

Perhaps, then, worrying about the future impact of climate change is a luxury affordable only to those who are lucky enough not to be living in that future already.

While I agree with Schneider-Mayerson about the need to raise these kinds of questions in relation to texts written from a narrow, privileged perspective that may elide climate justice, I do have a problem with his seeming acquiescence in and perpetuation of the dominance of such perspectives. After all, he may find fault with Rich and Robinson, but by devoting the bulk of his article to a close reading of their novels, he ironically ends up cementing their reputation as writers of climate change fiction worthy of critical attention and misses an opportunity to foreground literary texts (whether American or not) that do engage with issues of climate justice. Insofar as he presents himself as a detached outside observer, he fails to acknowledge and take responsibility for his own de facto role as a gatekeeper, an active agent in the constitution of what counts as valuable or "serious" climate change literature. By limiting the scope of his inquiry to fiction produced by white, male Americans, he himself seems to me to fall into the trap of "superpower parochialism" and "imperial narcissism" that he proposes as a possible explanation for the alleged paucity of literature dealing with issues of climate justice (960).

As a scholar and teacher of such literature myself, I make a conscious effort to assign and study often noncanonical texts from around the world that call attention to the plight of those most vulnerable to global warming. For example, I devote a class to *The Swan Book*, a novel by the Indigenous Australian writer Alexis Wright that illustrates the devastating impact of climate change on the Aboriginal community, which I ask students to read alongside Klein's essay about the production of sacrifice zones and disposable people through various othering mechanisms. I also assign "An Athabasca Story," a short story about oil extraction in the Alberta tar sands (an iconic sacrifice zone) by the Indigenous Canadian writer Warren Cariou that critiques the logic of settler-colonial petromodernity by highlighting its environmental and human cost. I teach this story alongside "Time Capsule Found on the Dead Planet," a breathtakingly ambitious flash fiction story by Margaret Atwood that traces the entire history of humanity. I try to show how Cariou's narrative breaks up the homogeneous "we" of Atwood's story, revealing heterogeneity and difference.

Another example of Indigenous climate change storytelling that articulates climate justice claims, and which I also like to discuss with my students, is a six-minute video poem called "Rise" by Kathy Jetñil-Kijiner and Aka Niviâna, Indigenous poets from the Marshall Islands and Greenland respectively. The poem results from an expedition undertaken by two islanders that connects their realities of melting glaciers and rising sea levels. Niviâna's way of life is disappearing as her country thaws, while the subsequent meltwater threatens Jetñil-Kijiner and her fellow Marshall Islanders thousands of miles away. Without mentioning any names, the two poets indict not human nature or people in general but "colonizing monsters" (00:03:38) for the violence, suffering, and pollution inflicted on their islands, of which the climate crisis is but the latest instance, and call on people to rise up in protest.

It seems to me that, as engaged scholars and teachers, we have a responsibility to amplify these kinds of unheard or scarcely heard voices, or, at the very minimum, to avoid perpetuating their silencing and thereby becoming actively complicit in climate injustice.

Works Cited

Atwood, Margaret. "Time Capsule Found on the Dead Planet." *The Guardian*, 26 Sept. 2009, www.theguardian.com/books/2009/sep/26/margaret-atwood-mini-science-fiction.

Cariou, Warren. "An Athabasca Story." *Lake*, vol. 7, 2012, pp. 70–75.

Crutzen, Paul J., and Eugene F. Stoermer. "The 'Anthropocene.'" *IGBP Newsletter*, no. 41, 2000, pp. 17–18.

Davis, Heather, and Zoe Todd. "On the Importance of a Date, or Decolonizing the Anthropocene." *ACME: An International Journal for Critical Geographies*, vol. 16, no. 4, 2017, pp. 761–80.

First Reformed. Directed by Paul Schrader, Lions Gate Entertainment, 2018.

Haraway, Donna, et al. "Anthropologists Are Talking—about the Anthropocene." *Ethnos*, vol. 81, no. 3, 2015, pp. 535–64.

Jetñil-Kijiner, Kathy, and Aka Niviâna. "Rise." *350.org*, 350.org/rise-from-one-island-to-another/.

Klein, Naomi. "Let Them Drown: The Violence of Othering in a Warming World." *London Review of Books*, vol. 38, no. 11, 2 June 2016, www.lrb.co.uk/the-paper/v38/n11/naomi-klein/let-them-drown.

Malm, Andreas. *Fossil Capital: The Rise of Steam Power and the Roots of Global Warming*. Verso Books, 2016.

McEwan, Ian. *Solar*. Random House, 2010.

Mehnert, Antonia. *Climate Change Fictions: Representations of Global Warming in American Literature*. Palgrave Macmillan, 2016.

Moore, Jason W. *Capitalism in the Web of Life: Ecology and the Accumulation of Capital*. Verso Books, 2015.

Nixon, Rob. "The Anthropocene: The Promise and Pitfalls of an Epochal Idea." 6 Nov. 2014. *Edge Effects*, 12 Oct. 2019, edgeeffects.net/anthropocene-promise-and-pitfalls/.

———. *Slow Violence and the Environmentalism of the Poor.* Harvard UP, 2011.
Schneider-Mayerson, Matthew. "Whose Odds? The Absence of Climate Justice in American Climate Fiction Novels." *ISLE: Interdisciplinary Studies in Literature and Environment*, vol. 26, no. 4, 2019, pp. 944–67.
Simpson, Helen. *In-Flight Entertainment.* Vintage Books, 2011.
Take Shelter. Directed by Jeff Nichols, Hydraulx Entertainment, 2011.
Tsing, Anna L. *The Mushroom at the End of the World: On the Possibility of Life in Capitalist Ruins.* Princeton UP, 2015.
Whyte, Kyle. "Indigenous Climate Change Studies: Indigenizing Futures, Decolonizing the Anthropocene." *English Languages Notes*, vol. 55, nos. 1–2, 2017, pp. 153–62.
———. "Is It Colonial Déjà Vu? Indigenous Peoples and Climate Injustice." *Humanities for the Environment: Integrating Knowledge, Forging New Constellations of Practice*, edited by Joni Adamson and Michael Davis, Routledge, 2017, pp. 88–105.
Wright, Alexis. *The Swan Book.* Atria Books, 2016.

Andrew Hageman

Engaging Students and Global Weirding

I started drafting this essay outside my campus office. Inside my office, the carpet was soaked and a ceiling tile hung precariously, like a water balloon made of mineral fiber. Aging HVAC infrastructure couldn't keep up with a week of high temperatures that were fifteen degrees Fahrenheit above average, so my workplace had to shift. I finished revising the essay after driving across South Dakota and Minnesota through a massive blanket of forest fire smoke. Global weirding.

This essay articulates opportunities that exploring weird fiction in courses on the literature of climate change and related subjects can create. The objectives, activities, assignments, and outcomes that follow emerge from experience designing and delivering two distinct iterations of a global-weirding-themed literature course. The first iteration was a capstone seminar for advanced undergraduates—mostly, though not exclusively, English majors—a few years before the pandemic. The second iteration was an intensive seminar open to all first-year students, spanning the month of September 2020, taught face-to-face with masks, physical distancing, and sanitizing protocols in the classroom. In their differences, these iterations present a dialectical argument for the startling productivity of weird fiction in climate change literature classrooms. Before COVID-19, weird literary fiction and critical theory texts provoked students to imagine portals out of existing ideological spaces and into relatively open spaces where they could experiment with alternative, often strange, perspectives. In the midst of COVID-19 and planetwide protests following the killing of George Floyd, those texts continued to build students' capacities to use the time of painfully weird uncertainties to analyze existential threats and imagine radical changes for the good of people and the biosphere as a whole.

Texts

For the purposes of course designs and this essay I leave the generic definition of *weird literature* quite open. This constellation of work was initially inspired by novels often classified as the New Weird (Jeff Vander-Meer's *Annihilation* serves as a fundamental example). However, my central pedagogical aim is the collective analysis of how literature that swirls together the speculative, the strange, and the surreal empowers readers with new paradigms through which to grasp climate change and conceptualize future change. To that end, I discuss two such works below in terms of some critical elements they contribute to the literature of climate change, then present a selection of classroom activities that readers may wish to consider using in their own courses.

In the advanced seminar, David Mitchell's *Cloud Atlas* prompted the most sustained discussions and was overwhelmingly the students' favorite read. The novel has a far-future section at its center; five other sections, set in different time periods, are divided so that half of each section is read before the center and half after. These split sections follow chronological order, so readers go from the farthest past to the far future and then return through the same temporal strata to the farthest past. Mitchell's formal innovation creates a narrative through line that spans multiple human generations as well as practices and events that change the planet's climate in chronic and acute ways. Akin to the intercalary chapters of John Steinbeck's *The Grapes of Wrath*, the section shifts in *Cloud Atlas* disrupt smooth narrative flows to jolt readers into a range of disparate perspectives. Both novels enact productive heteroglossia to collect points of view that cut across cultures, classes, species, and many more categories that readers might bring to them. *Cloud Atlas*, in particular, locates the reader's present on a continuum that tracks with climate change as an aggregation of past and continuing actions that reverberate and intensify, thereby increasingly haunting the future. Many students responded to the novel by shifting their perception of the characters—and of themselves—such that identity became grounded not in where people lived but in when they lived. Students also invoked climate change to interpret the comet-shaped birthmark borne by a different character in each of the novel's sections. The birthmark transcends individuals and signals an abiding love even as those who love must navigate the troubled planet and power structures built in the past.

The novel I found most productive in both the advanced course and the first-year course was Johanna Sinisalo's *Troll: A Love Story*. Seemingly set very near 2000, when the work was first published, *Troll* tells the tale of a young man in Helsinki who rescues an adolescent troll near his apartment one night and develops an incredibly complicated relationship with this fellow earthling. Sinisalo syncs up climate change with queer subjectivities and extractivist attitudes toward nonhuman beings, including humans who are systematically dehumanized so as to be made a resource. The appearance of this troll in Helsinki is depicted subtly as one instance of the mass trend in nonhuman refugees displaced by climate change and habitat devastation. Multiple chapters leverage a mapping motif to spark readers to inhabit the same geographical area vicariously through queer human and other subjectivities. These passages deconceal the structures of oppression made concrete through design, architecture, infrastructure, and so on by showing objects as observed or used by diverse beings. *Troll*'s narrative chapters are intercut with chapters that excerpt reference texts and news media, most of which are fictional creations by Sinisalo (though the author remarked in an interview that she has received multiple inquiries over the years about her "sources"). These intercalations juxtapose traditional Finnish lore and mythology with science in a way that respects both modes while highlighting the weirdness of each as well as their historical complicity in planetary damage. Through the entangled human-troll plot and paratexts, Sinisalo creates a weird literary fiction dynamic that would pair wonderfully with Chanda Prescod-Weinstein's *The Disordered Cosmos*. Discussions might focus on both novels' treatment of love, sex, gender, race, species, space-time, climate change, and political economy.

I construct my syllabus with the understanding that literature of climate change courses inherently call for a diverse, equal, and inclusive collectivity of writer and character identities. The responsibilities for carbon emissions as well as the experiences of the warming they foment are unevenly distributed. Similarly, the capacities to design ways to reduce carbon emissions and adapt to the changes that are already inevitable differ widely across cultures, economic practices, knowledge sets, and approaches to architecture and infrastructure. For these reasons, a syllabus should engage a range of perspectives. Some texts I have used to this purpose include Michel Faber's *Under the Skin*, Nnedi Okorafor's *Lagoon*, N. K. Jemisin's "The Ones Who Stay and Fight," Laurie Penny's *Everything Belongs to the Future*, Octavia E. Butler's *Parable of the Sower*, Stephen Graham Jones's *The Only Good Indians*, China Miéville's "Covehithe," and Berit Ellingsen's *Vessel and Solsvart*.

Disaggregating the Chapter

The classroom activities I describe below have shown that teaching the literature of climate change is an ideal platform for developing students' close reading acumen. For early undergraduates or nonmajors, instructors can craft close reading assignments that warmly invite folks with diverse literary study preparation and learning styles. Assignments for advanced students can help expand their close reading skills and align their literary analysis with competencies they will need as professionals, community members, and residents of a changing earth.

One assignment I designed and have adapted to many of my literature courses, whether ecocritical, survey, or genre-based, is Disaggregating the Chapter. I create student teams and assign each one a different chapter from the day's reading to disaggregate. In the prepandemic classroom, each team would move to a wall-mounted dry-erase board and work with colored markers. In the pandemic-protocol classroom, teams gathered in physically distanced clusters or by video meeting and built slides to share. I distribute a common set of data categories, such as geographical movements, time covered, and pace of action, to the teams and allow at least ten minutes for teams to discuss how they will identify a passage in the chapter that fits a given category and how they want to present the overall chapter findings visually. Both parts of this discussion are key. Identification gets at the heart of close reading since students will need to collaborate on labeling literary content or form with a particular category tag. A category like "time covered" may initially seem straightforward, but in application students will tune in to flashes backward and forward and to invocations of trauma that point to moments in the past or, through E. Ann Kaplan's climate change concept "pretraumatic stress disorder," in the future. Throw in an excavation scene, perhaps one that turns up an alien artifact that traveled billions of years before landing on Earth, and time gets weirder and weirder. Metaphor, metonymy, prolepsis, free indirect discourse, and other close reading tools come into play. Diving into the details of tagging leads to student discussions of how to present the data visually. For example, a standard time line, which depicts a straightforward duration of events, might not effectively convey the temporal complexities that abound in a lot of weird fiction.

After the category preparation above, student teams transition into searching the text itself to produce their data sets. This phase of the assignment should be allocated a substantial portion of the session (I allowed half of a ninety-minute period). I interact intensively with all teams during this phase to react, reinforce, and redirect on a case-by-case basis. Finally, after the teams have created visual representations for all categories, I ask each team to examine another team's work, raise questions based on looking at the data, and forge some hypotheses that might answer these questions. The next phase, a full-class discussion of each chapter, tends to reach into a subsequent class session. Be sure to capture the teams' board work or slides and initial reactions to their productions by taking photos of the boards or saving slide decks and sharing these with students on the course management system.

In the advanced seminar, students completed this assignment on mid-novel chapters of Michel Faber's *Under the Skin*. Categories included characters present, geographical movements, time covered, weather, mood, pace of action, pace of prose, point-of-view patterns and breaks, the chapter's first paragraph, and the chapter's last paragraph. The weather category as a focus, and then in conversation with other categories, produced the highest-impact insights. The protagonist, Isserley, an extraterrestrial alien who drives around rural Scotland to pick up hitchhikers who are then processed as meat to export to her home planet, tends to contemplate weather phenomena in sustained passages. Students listed the numerous weather passages within that category and observed that Isserley was often the lone character in these moments; they also noted several point-of-view shifts between Isserley and Amlis Vess, the wealthy extraterrestrial heir to this meat corporation, in weather-intensive passages. After they pinpointed these convergences of data categories in the full-class discussion, we all opened our books and conducted collaborative close reading of particular instances of convergence.

Collectively, we worked through passages to some agreement that Amlis Vess's point of view as an alien in his first days on Earth contrasted with Isserley's partly assimilated point of view. Whereas Isserley pulls human readers sensuously into weather, Amlis Vess marvels at the planet's climate. The bare fact of water in the atmosphere strikes him as radically weird. Rain and snow pass from hitchhikers' associations with traffic hazard and discomfort through Isserley's sublime aesthetic appreciation to Amlis Vess's distinct awareness of climate and climate change as he contrasts Earth with his and Isserley's home planet. Through the process of disaggregating the chapter and experimenting with data sets in dialogue, this assignment, a sort of nanoversion of Franco Moretti's distant reading, enables students to ask questions about climate change in this weird novel that they otherwise might not think to ask, and it guides them to construct and test hypothetical answers through close reading. In this way, the assignment models how to engage with climate change by fusing the form and content of literary narrative with algorithmic data science.

Coimagining Weirdness

I will share one more illustrative classroom exercise in compact form. To activate nonvisual human senses and achieve more universal design, I developed a sonic assignment for discussing Nnedi Okorafor's *Lagoon*. Okorafor writes in multiple sensory registers, including moments of synesthesia, so the concept of the assignment maximizes students' awareness of this feature of the novel. First, I ask students to select a passage from the day's reading they consider connected to climate change—this works well in pairs. Then I ask them to mark up the passage with an ear toward how they would use sound design to produce an audiobook that uses sensory immersion to highlight the ecological aspect of *Lagoon*. As weird fiction, *Lagoon* features intricate layers in the presences and absences of sounds from human bodies and technologies, from alien bodies and technologies, and from nonhuman terrestrial objects. Students reacted to the assignment as a surprising discovery of a sensory wavelength that had shaped their readings subtly yet significantly. For first-year and nonmajor students, this exercise transcends the "find the hidden symbols" approach to "decoding" literature that many bring to college. Reading becomes the work of coimagining. This exercise also moved students to consider how they could accurately and responsibly research the soundscape of Lagos, Nigeria, where *Lagoon* is largely set. We discussed the intersection of representing places, peoples, and cultures with the fostering of empathy through multiple sensory experiences of being.

The exercise pairs well in a global weirding and literature of climate change context with Björk's *Biophilia* concert film and Cosmo Sheldrake's *Wake Up Calls* album. Sheldrake's descriptions of his work speak clearly about the sonic recording and registering of climate change in any given place by documenting the shifting presences and absences of particular sounds, and his work will help students tune in to the role sound plays in climate change literature and their own lived experiences.

Expedition Journal

The last assignment I will describe can span as much of a term as the instructor desires. What I call the expedition journal activates students' creative modes and deepens their connections to place (thereby helping first-year students become attached to their new home and zone of learning). My idea for the expedition journal was sparked by Jeff VanderMeer's work. Expeditions and journals are core to his Southern Reach trilogy, and VanderMeer has remarked that many of the trilogy's weird descriptions of Area X, where the action is set, came straight from the pages of his own hiking journals (Featured speaker keynote address). Students thus informed must tangle with the fact that the weirdest material in the story hews closely to actually experienced places. The university where we learn together, in Decorah, Iowa, acknowledged as Sioux, Sauk and Meskwaki, and Wahpeton land, sits in a meteor crater valley where fossils of seven-foot-long sea scorpions (*Pentecopterus decorahensis*) were exposed by the historic floods of 2008 that were shaped by climate change. These and other elements of our specific location's human and nonhuman history enabled students to stretch their imaginations and to explore how weird fiction can help folks perceive places and processes in new and productive ways.

In the most recent iteration, during the pandemic, I aligned the frequency and focus of journal entries with syllabus readings and the intensive nature of the 3.5-week course, but prompts can be customized for different locations and modes of learning.

Week 1: Liminal Zones. I assigned minimum time frames for students to spend in two different liminal zones of their choosing on or near our residential campus: one where water meets land and one where two landscapes meet. I provided questions to serve as prompts for journal entries: What productive and destructive dynamics happen at the border and as you move slowly away from it in either direction? What is alive in the zone and what is not? How are these elements interacting? Students were asked to use their experience to write the journal entry as part of a short story.

Week 2: In the Dark. The second journal assignment asked students to select—prioritizing personal safety—an outdoor location to explore at night in minimal artificial light, dividing their time between sedentary and moving experiences. I encouraged students to consider how various elements of their surroundings differed from their daytime, daylit forms. Gazing upward, whether the sky was cloud-shrouded or star-stippled, they might reflect on nighttime, deep time, dark matter, and climate change. For this journal, students were invited to write a weird poem inspired by their explorations. It is worth emphasizing that nighttime darkness can pose various safety challenges and that students should be encouraged to explore in groups, take along flashlights and phones, and choose areas that instructors have deemed likely to be safe.

Week 3: One Map or Many? This journal entry combined cartography with prose writing as students were asked to choose an area no larger than a city block and make between three and five maps of it from various human and nonhuman perspectives—such as that of a winter wind from the north or a raindrop driven by an east wind. As inspiration for this assignment, an excerpt from Ian Bogost's *Alien Phenomenology* that considers a block of tofu's experience of swimming inside a plastic package can crack open room for students to get wonderfully weird (7). This journal submission took the form of hand- or digitally drawn visuals accompanied by written descriptions that invited readers inside the world of each character.

The expedition journal assignment opened the ideological space for students to reframe, similarly to the authors we had read in the course, the places they experience daily as an accretion of climate and other changes over vast spans of time. The exercise helped them become more flexible in imagining climate change futures and the urgency human beings must bring to shaping a future earth if we want ourselves included in it.

All the objectives, text selections, and practices I present here aim to challenge students and instructors in our collective work engaging with climate change through literature. Weird literature embraces surreal aesthetics, unsettling montage, and disorienting layering of scopes and scales. In this time of carbon emissions crux, leveraging the strange forces in these texts and their readers can bring energy and innovation to the manifold endeavor of our future on Earth.

Works Cited

Biophilia Live. Performance by Björk, directed and edited by Peter Strickland and Nick Fenton, Cinema Purgatorio, 2014.

Bogost, Ian. *Alien Phenomenology; or, What It's Like to Be a Thing.* U of Minnesota P, 2012.

Butler, Octavia E. *Parable of the Sower.* Four Walls Eight Windows, 1993.

Ellingsen, Berit. *Vessel and Solsvart.* Snuggly Books, 2017.

Faber, Michel. *Under the Skin.* Harcourt, 2000.

Jemisin, N. K. "The Ones Who Stay and Fight." *Lightspeed*, no. 116, Jan. 2020, www.lightspeedmagazine.com/fiction/the-ones-who-stay-and-fight/.

Jones, Stephen Graham. *The Only Good Indians.* Saga Press, 2020.

Kaplan, E. Ann. *Climate Trauma: Foreseeing the Future in Dystopian Film and Fiction.* Rutgers UP, 2015.

Miéville, China. "Covehithe." *Three Moments of an Explosion: Stories*, by Miéville, Del Rey, 2016, pp. 299–311.

Mitchell, David. *Cloud Atlas.* Random House, 2004.

Moretti, Franco. *Graphs, Maps, Trees: Abstract Models for a Literary History.* Verso Books, 2007.

Okorafor, Nnedi. *Lagoon.* Hodder and Stoughton, 2014.

Penny, Laurie. *Everything Belongs to the Future.* Macmillan Publishers, 2016.

Prescod-Weinstein, Chanda. *The Disordered Cosmos: A Journey into Dark Matter, Spacetime, and Dreams Deferred.* Bold Type Books, 2021.

Sheldrake, Cosmo. *Wake Up Calls.* Tardigrade Records, 2020.

Sinisalo, Johanna. Interview with Jeff VanderMeer. WorldCon 75, 10 Aug. 2017, Helsinki.

———. *Troll: A Love Story.* Translated by Herbert Lomas, Grove Press, 2004.

VanderMeer, Jeff. *Annihilation.* Farrar, Straus and Giroux, 2014.

———. Featured speaker keynote address. Sigma Tau Delta Annual Convention, 31 Mar. 2017, Louisville, KY.

April Anson

Toward a Critical Environmental Justice Pedagogy

In his *Book of Delights*, the poet Ross Gay ponders a question from one of his students, who has proposed, as a vision for the classroom she herself might lead one day, "What if we joined our wildernesses together?" Taking up his student's metaphor, Gay asks:

> Is sorrow the true wild?
> And if it is—and if we join them—your wild to mine—what's that?
> . . . .
> I'm saying: What if that is joy? (49–50)

In the context of the climate crisis, what would happen "if we joined our wildernesses together?" Gay asks parenthetically, "[H]ave I made the metaphor clear?"—acknowledging the ambivalence of environmental metaphors for the human experience (49). Surely, there is an intense, enduring power in ecological analogies for human emotions and a special value in the metonymic recognition of our own annihilation in the existential threat that is climate chaos. Still, Gay's question "Is sorrow the true wild?" whispers the profound political power of the wilderness concept. Especially when spoken in English, *wilderness* signifies a mythology of nature used to justify colonial land theft, genocide, slavery, extraction economies, and the pervasive, generational grief of dispossession that accompanies settler colonialism. If, following William Cronon, the wilderness metaphor is nothing but trouble, what does that mean for our sorrow for it, for us who are it? Even when they remain opaque or troubled, our sorrows and joys rarely feel metaphorical. They are visceral and fickle, just as historically fraught, contested, and full of culturally divergent meanings as the wilderness metaphor itself. Still, our sorrows do sometimes and, "somehow, meet" (Gay 49) when we read, discuss, and otherwise think literarily with a changing climate. The literature of climate change challenges us to rethink the relationship between history and present, fiction and fact, a student's practice and a teacher's instruction. This essay articulates a set of strategies to move us toward a critical environmental justice pedagogy for teaching the literature of climate change.

Climate literacy, especially in its literary form, speaks not just to the difficulty of envisioning what climate science says without the aid of storytelling tools like metaphor but also to the ways that literary methods, and the practice of the literary imagination itself, are absolutely essential to considerations of climate change. Metaphor is one of the most frequently recommended strategies for communicating climate science (see Deignan et al.; Armstrong et al.; Niebert and Gropengiesser; "Climate" [*Smithsonian*]). The National Aeronautics and Space Administration uses the metaphor of waves in the bathtub to explain sea-level rise (Willis); the Metaphor Project emphasizes the rhetorical power of words like *clean* instead of *green* and *rules* instead of *regulations* ("Grow"); and K–12 teachers use long lists of metaphors in a single lesson plan ("Climate" [*WildBC*]). This broad attention to metaphor testifies to the art embedded in discussions of climate change as well as to the vitality of literary methods for addressing the profoundly personal and inevitably political past, present, and future—or potential futures—of climate chaos. As with the wilderness metaphor, environmental science and stories writ large demand we discuss the devastation caused by white supremacy and the colonial violence that formed and continues to fuel the extraction economics most responsible for global heating.

Our pedagogies must be oriented likewise. We have an ethical and material obligation to center questions of justice and inequality in our literary explorations of environmental issues. Inspired by David Pellow and Robert Brulle's formative focus on intersectionality, scalability, and commitment to deepening direct democracy, and named in homage to Pellow's *What Is Critical Environmental Justice?*, a critical environmental justice pedagogy (CEJP) emerges from my observation that best practices for teaching climate change deeply resonate with recommended strategies for teaching race and ethnicity and echo high-engagement teaching, especially for second language learners. CEJP is drawn from my almost twenty years of experience teaching environmental justice and climate change literature with high school, undergraduate, graduate, and nontraditional students, as well as other teachers, in public and elite institutions. CEJP synthesizes best practices for active learning, cultivating diversity and inclusion ("Resources"); the Sheltered Instruction Observation Protocol for teaching second language learners ("What"); the interdisciplinary college programs Writing in the Disciplines (writinginthedisciplines .com/) and Writing across the Curriculum (Dalporto); and the anti-racist and climate pedagogy research detailed below. When we acknowledge that social injustice is the basic unit of environmental injustice, the convergence of these strategies makes supreme sense.

General principles for the intersection of these pedagogical fields stress writing, inquiry, collaboration, interdisciplinarity, and reading in every session. They suggest structured discussion with clear prompts and objectives, providing key vocabulary through collaboratively developed and open access glossaries, offering multiple models of assignments, incorporating personal storytelling, and using real-world and local examples of climate chaos and efforts to counteract it. CEPJ is possible in active learning strategies like Socratic fishbowls, think-pair-shares, give-one/get-one, reciprocal teaching, and facilitating student interaction every fifteen minutes (see Gifkins). Specifically, classes on climate change and the arts tend to center the real-world stakes of artistic production, the emotional impacts of climate chaos, and the desire for solutions to environmental crisis. Below, I use these three themes to structure a schematic of CEJP strategies, guided by anti-racist pedagogical criteria, that urges educators to encourage reflexivity, prepare for and welcome difficulty, meet students where they are, engage affective and embodied dimensions of learning, and build learning communities (Thurber et al.; see also Brosheim-Black and Saraigianides; "Anti-Racist Pedagogy"; "Anti-Racism").

Real and Fictional Climates

We can locate a changing climate in chronicles of weather, water quality, and wildfires—but also through a character's wonder, fear, laments, and confusions, which resonate with our own. Climate change blurs the boundaries between the real and the surreal, between fact and fiction, and can lead students to treat "fictions as unmediated sources of information" (Garrard 123). Therefore, we must include a diversity of sources, perspectives, genres, and historical periods. Rhetoric around current events, for example, can and should be treated as text to analyze. In other words, the methods of literary analysis equip us to navigate the facts and fictions of climate change.

CEPJ strategies synthesize the tools of literary study with best practices for teaching climate change. CEJP follows the collection *Teaching Climate Change in the Humanities* (Siperstein et al.), which recommends normalizing struggle with difficult concepts, topics, and emotions. Shawn Wilson's work can inform a shift in a classroom's focus from unanswered questions to unquestioned answers; similarly, Anthony Lioi helps prepare educators to remind students that concepts often taken for granted, like *nature* or *citizen*, can be contested. Ann E. Green, Wiley Davi, and Olivia Giannetta offer strategies for incorporating place-based teaching through service learning, public storytelling, case studies, collaboratively defined research methods, revisions to historic markers, or *Wikipedia* edit-a-thons. The collection *Pipeline Pedagogy* (Banschbach and Rich), David Orr's "pedagogy of transition," Zoe Todd's "watershed level thinking," and la paperson's "ghetto land pedagogy" prove that pedagogy must be grounded in material realities and lived experiences. In preparing to discuss particular subject positions in relation to systems of power, instructors can benefit from taking the "Anti-Racist Educator Questionnaire" (Stamborski et al.) and using self-reflective practices like backward planning, coauthored glossaries, and treating themselves as texts, drawing on the work of Sarah Ensor, Teresa Newberry and Octaviana V. Trujillo, and Dolores Calderon. Ensor argues that collaborative learning helps students and teachers respond to the precarity of a changing ecological and institutional climate. Newberry and Trujillo show that talking about climate change entails the recognition of vital differences in institutional environments, lived experiences, traumas, and emotional responses, echoing Calderon's attention to the settler grammars of academic institutional settings. In all, the analytical tools of literary analysis can clarify the intimacies often lost in climate education—namely, what a changing climate means to and in our locales, bodies, and biases.

Affect

As Sarah Jaquette Ray discusses in *A Field Guide to Climate Anxiety*, most educators are not trained to deal with the profound psychological tumult that often accompanies the overwhelming facts, traumatic lived experiences, and politically charged realities of climate change and its causes. CEJP incorporates trauma-informed instruction that issues content warnings, plans time for recovery, and encourages students to self-monitor and adjust accordingly (see Bryan; Zingarelli-Sweet). Instructors should leave space for grief and worry to be shared or expressed privately in a journal or other form of response, like the Climate Change Worry Scale (Stewart). Using a gradation of mild to severe, classes can brainstorm a list of climate emotions that can be revisited in lectures and activities (Wray, "Why"), including solastalgia, the distress caused by the degradation of one's environment (Albrecht et al.). It may be helpful to acknowledge fear as a spectrum that is not just about "survival" (brown), reframe the binary of fear and guilt into the accountability and action of "prospective survivors" (Hemphill; Wray, "It"), and employ "Broken World Teaching" to regard failure as a given (Siperstein). These approaches pair well with an exposé on BP's carbon-footprint PR campaign or Shell's investment in climate disinformation. Instructors can follow the Indigenous, decolonial feminist scholar Dian Million's concept of "felt research," which conceptualizes feelings as theory and "culturally mediated knowledges, never solely individual" (61). Similarly, Kyle Whyte's bibliography of climate justice teaching materials and Ray's article "Climate Anxiety Is an Overwhelmingly White Phenomenon" are valuable related resources. In other words, hope is a feeling and a tool. In *Radical Hope*, Kevin Gannon says, "Teaching is a radical act of hope. It is an assertion of faith in a better future in an increasingly uncertain and fraught present. It is a commitment to that future even if we can't clearly discern its shape" (5). Still, hope can entail minimizing the terrifying realities of climate change. Contrast Greta Thunberg's viral speech "I don't want you to be hopeful" with the research that shows that hope is learnable and results in higher self-esteem and goal attainment (Thunberg; Fleming). Still, CEJP regards nihilism, denial, and even extremism in the classroom as harmful and misdirected but comprehensible as coping strategies (Haltinner et al.; Haltinner and Sarathchandra; Schmidt). Rather than disregard these responses, CEJP recommends *A Teacher's Guide on the Prevention of Violent Extremism*, published by UNESCO, as a vital resource regarding concerning opinions or behaviors.

Solutions

Student thrive when given solutions, especially when those solutions are presented as partial, emphasizing the diversity of work needed to address climate change. CEJP courses can model the diversity and possibility of solutions by assigning students to research a real-world (partial) solution or a local organization dedicated to the issue at hand, acknowledging repeatedly that there is no silver bullet and discussing student concerns. CEJP recommends using the search function on the websites maintained by the Climate Literacy and Energy Awareness Network (cleanet.org/index.html), the National Science Teaching Association (nsta.org/), and the National Center for Science Education (ncse.ngo/) as well as NASA's *Global Climate Change: Vital Signs of the Planet* website (climate.nasa.gov/), the National Oceanic and Atmospheric Administration's *Climate Education* website (noaa.gov/climate-education), and the University of California's *UC-CSU NXTerra* website (nxterra.orfaleacenter.ucsb.edu/).

Mitigating climate change will take all of us, doing what we can. I end the term on the themes of goals and accountability, using the Venn diagram designed by Ayana Elizabeth Johnson, cohost of *How to Save a Planet* ("Is Your Carbon Footprint BS?"). The diagram asks, "What are you good at? What is the work that needs doing? What brings you joy?" The three spheres intersect in the space where the questioner is poised to be most useful and fulfilled in the climate movement. Sharing individual answers in class never fails to inspire.

CEJP is meant to support critical environmental justice goals through a set of pedagogical practices that can be incorporated and modified to suit individual teaching styles and student populations. There is no such thing as a perfectly transferrable, universal pedagogy. Even environmental science shows that diversity is the foundation for flourishing. When we witness shared sorrows and share radically divergent wild spaces, we confront our uneven places within the continuing impacts of colonial capitalism. In this vital diversity, inside and outside the classroom, there is a joining together that feels, indeed, quite like joy.

Works Cited

Albrecht, Glenn, et al. "Solastalgia: The Distress Caused by Environmental Change." *Australasian Psychiatry*, vol. 15, no. 1, supp. 1, Feb. 2007, pp. S95–98, https://doi.org/10.1080/10398560701701288.

"Anti-Racism and Allyship in the Classroom." docs.google.com/document/d/1or-vfugiIbQcgCA9WEeq1cL-Nu03OglyqHmk8dv-JV8/.

"Anti-Racist Pedagogy in Action." *Columbia Center for Teaching and Learning*, ctl.columbia.edu/resources-and-technology/resources/anti-racist-pedagogy/.

Armstrong, Anne K., et al. "Using Metaphor and Analogy in Climate Change Communication." *Communicating Climate Change: A Guide for Educators*, by Armstrong et al., Cornell UP, 2018, pp. 70–74.

Banschbach, Valerie, and Jessica L. Rich, editors. *Pipeline Pedagogy: Teaching about Energy and Environmental Justice Contestations.* Springer, 2021.

Brosheim-Black, Carlin, and Sophia Tatiana Sarigianides. *Letting Go of Literary Whiteness: Anti-Racist Literature Instruction for White Students.* Teachers College Press, 2019.

brown, adrienne maree. *We Will Not Cancel Us and Other Dreams of Transformative Justice.* AK Press, 2020.

Bryan, Audrey. "Affective Pedagogies: Foregrounding Emotion in Climate Change Education." *Policy and Practice: A Development Education Review*, no. 30, spring 2020, www.developmenteducationreview.com/issue/issue-30/affective-pedagogies-foregrounding-emotion-climate-change-education.

Calderon, Dolores. "Uncovering Settler Grammars in Curriculum." *Educational Studies: A Journal the American Educational Studies Association*, vol. 50, no. 4, July 2014, pp. 313–38.

"Climate Change Metaphors." *Smithsonian Ocean*, Smithsonian National Museum of Natural History, ocean.si.edu/educators-corner/climate-change-metaphors.

"Climate Change Metaphors." WildBC, www.tigurl.org/images/tiged/docs/activities/615.pdf.

Cronon, William. "The Trouble with the Wilderness; or, Getting Back to the Wrong Nature." *Environmental History*, vol. 1, no. 1, 1996, pp. 7–28.

Dalporto, Deva. "Writing across the Curriculum." *We Are Teachers*, 25 June 2013, www.weareteachers.com/writing-across-the-curriculum-what-how-and-why/.

Deignan, Alice, et al. "Metaphors of Climate Science in Three Genres: Research Articles, Educational Texts, and Secondary School Student Talk." *Applied Linguistics*, vol. 40, no. 2, 2019, pp. 379–403.

Ensor, Sarah. "Relative Strangers: Contracting Kinship in the Queer-Ecology Classroom." *American Literature*, vol. 89, no. 2, 2017, pp. 279–304.

Fleming, Nora. "In Schools, Finding Hope at a Hopeless Time." *Edutopia*, 26 Mar. 2021, www.edutopia.org/article/schools-finding-hope-hopeless-time.

Gannon, Kevin M. *Radical Hope: A Teaching Manifesto.* West Virginia UP, 2020.

Garrard, Greg. "In-Flight Behavior: Teaching Climate Change Literature in First-Year Intro English." Siperstein et al., pp. 118–25.

Gay, Ross. *The Book of Delights.* Algonquin Books, 2019.

Gifkins, Jess. "What Is 'Active Learning' and Why Is It Important?" *E-International Relations*, 15 Oct. 2015, www.e-ir.info/2015/10/08/what-is-active-learning-and-why-is-it-important/.

Green, Ann E., et al., editors. *Social Justice Writing Assignments: Toward a Politics of Location.* Special issue of *Prompt: A Journal of Academic Writing Assignments.* Vol. 5, no. 1, 2021.

"Grow Your Impact on Climate Action Framing." *The Metaphor Project*, 30 Jan. 2020, metaphorproject.org/wp-content/uploads/2021/04/Grow_Your_Impact_4f.pdf.

Haltinner, Kristin, and Dilshani Sarathchandra. "Climate Change Skepticism as Psychological Coping Strategy." *Sociology Compass*, vol. 12, no. 65, 2018, pp. 1–21.

Haltinner, Kristin, et al. "Feeling Skeptical: Worry, Dread, and Support for Environmental Policy among Climate Change Skeptics." *Emotion, Space, and Society*, vol. 39, May 2021, https://doi.org/10.1016/j.emospa.2021.100790.

Hemphill, Prentis. "Letting Go of Innocence." *Prentis Hemphill*, 5 July 2019, prentishemphill.com/selectwritings/2019/7/5/letting-go-of-innocence.

"Is Your Carbon Footprint BS?" *How to Save a Planet*, 18 Mar. 2021, gimlet media.com/shows/howtosaveaplanet/xjh53gn/is-your-carbon-footprint-bs.

la paperson. "A Ghetto Land Pedagogy: An Antidote for Settler Environmentalism." *Environmental Education Research*, vol. 20, no. 1, 2014, pp. 115–30.

Lioi, Anthony. "Teaching Earth." *Transformations*, vol. 21, no. 1, 2010, pp. 14–22.

Million, Dian. "Felt Theory: An Indigenous Feminist Approach to Affect and History." *Wicazo Sa Review*, vol. 24, no. 2, fall 2009, pp. 53–76.

Newberry, Teresa, and Octaviana V. Trujillo. "Decolonizing Education through Transdisciplinary Approaches to Climate Change Education." *Indigenous Studies and Decolonizing Stories in Education*, edited by Linda Tuhiwai Smith et al., Routledge, 2019, pp. 204–14.

Niebert, Kai, and Harald Gropengiesser. "Understanding and Communicating Climate Change in Metaphors." *Environmental Education Research*, vol. 19, no. 3, 2012, pp. 282–302.

Orr, David. "The Pedagogy of Transition." *Resilience*, 25 Mar. 2021, resilience.org/stories/2021-05-25/the-pedagogy-of-transition-educating-for-the-future-we-want/.

Pellow, David Naguib. *What Is Critical Environmental Justice?* Polity, 2018.

Pellow, David Naguib, and Robert J. Brulle. "Poisoning the Planet: The Struggle for Environmental Justice." *Contexts*, vol. 6, no. 1, pp. 37–41.

Ray, Sarah Jaquette. "Climate Anxiety Is an Overwhelmingly White Phenomenon." *Scientific American*, 21 Mar. 2020, www.scientificamerican.com/article/the-unbearable-whiteness-of-climate-anxiety/.

———. *A Field Guide to Climate Anxiety: How to Keep Your Cool on a Warming Planet.* U of California P, 2020.

"Resources." *AVID*, 2023, avid.org/resources.

Schmidt, Charles W. "A Closer Look at Climate Change Skepticism." *Environmental Health Perspectives*, vol. 118, no. 12, 1 Dec. 2010, ehp.niehs.nih.gov/doi/full/10.1289/ehp.118-a536.

Siperstein, Stephen. "Broken World Teaching." *Hybrid Pedagogy*, 11 May 2021, hybridpedagogy.org/broken-world-teaching/.

Siperstein, Stephen, et al., editors. *Teaching Climate Change in the Humanities.* Taylor and Francis, 2016.

Stamborski, Anna, et al. "Anti-Racist Educator Questionnaire and Rubric." docs.google.com/document/u/0/d/1OT1-wV7ulYFpxQ3HAoo-7IID5Ap6s-MNEYwe3RRJGBo/.

Stewart, Alan E. "Psychometric Properties of the Climate Change Worry Scale."
The Psychological Impacts of Global Climate Change, special issue of *Environ-
mental Research and Public Health*, vol. 18, no. 2, 2021, article no. 494.
A Teacher's Guide on the Prevention of Violent Extremism. United Nations
Educational, Scientific, and Cultural Organization, 2016, en.unesco.org/
sites/default/files/lala_0.pdf.
Thunberg, Greta. "Greta Thunberg's World Economic Forum 2019 Special
Address." *Open Transcripts*, 24 Jan. 2019, opentranscripts.org/transcript/
greta-thunberg-world-economic-forum-2019/.
Thurber, Annie, et al. "Teaching Race: Pedagogy and Practice." Vanderbilt U
Center for Teaching, 2019, cft.vanderbilt.edu/teaching-race/.
Todd, Zoe. "From Classroom to River's Edge: Tending to Reciprocal Duties
beyond the Academy." *Aboriginal Policy Studies*, vol. 6, no. 1, 2016,
pp. 90–98.
"What Is the SIOP Model?" *Center for Applied Linguistics*, www.cal.org/siop/
about/index.html.
Whyte, Kyle. "Indigenous Climate Change and Climate Justice: Teaching
Materials and Advanced Bibliography." *U of Michigan School for Environment
and Sustainability*, kylewhyte.seas.umich.edu/climate-justice/.
Willis, Josh. "Waves in the Bathtub: Why Sea Level Rise Isn't Level At All." *Ask
NASA Climate*, 10 Jan. 2010, climate.nasa.gov/blog/239/waves-in-the
-bathtub/.
Wilson, Shawn. *Research Is Ceremony: Indigenous Research Methods*. Fernwood
Publishing, 2008.
Wray, Britt. "It Is Time to Step into the Role of the 'Prospective Survivor.'" *Gen
Dread*, 15 July 2021, gendread.substack.com/p/it-is-time-to-step-into-the-role.
———. "Why Activism Isn't *Really* the Cure for Eco-anxiety and Eco-grief."
Gen Dread, 5 Aug. 2020, gendread.substack.com/p/why-activism-isnt-really
-the-cure.
Zingarelli-Sweet, Desirae. "Keeping Up with . . . Trauma-Informed Pedagogy."
Association of College and Research Libraries, 22 June 2021, www.ala.org/
acrl/publications/keeping_up_with/trauma-informed-pedagogy.

**Ted Martinez

Changing Student Perceptions
through Climate Literature

Despite overwhelming scientific evidence, climate change still has a detection problem. That is, it is a phenomenon that some people perceive on the planet, in the news, or in their own lives, while others do not. The environmental philosopher Timothy Morton has labeled climate change a *hyperobject*, a term he coined to describe realities that are massive, nonlocal, and "sticky" in that they reveal themselves only through phenomena created by the interaction of other objects, a condition known as "interobjectivity" (72). As such, climate change defies humans' understanding of an object that can be easily seen, touched, or described and has a singular pinpoint location. Like the title character in the 1933 film adaptation of H. G. Wells's *The Invisible Man*, climate change cannot be detected except by measuring something associated with it, such as temperature, carbon dioxide in parts per million, or sea-level rise. These are the phenomena of climate change, and, like the bandages and clothing worn by the invisible man, they reveal the shape of the object. Without his suit, bandages, gloves, and sunglasses, the invisible man would indeed be invisible on screen. Similarly, climate change lacks a form, shape, or even location; it is revealed only through its phenomena, and this creates the detection problem. Some people can see climate change, while others cannot see it.

Although the scientific mechanism of climate change is well understood, many questions remain about how best to communicate the phenomenon and to engage citizens in creating societal and political transformation. Conventional scientific approaches to climate change education may be insufficient, and creative forms of climate change education and outreach may be more effective at changing perceptions (Corbett and Clark 3). Studies are beginning to show that environmental literature can change its readers' perceptions (Schneider-Mayerson 340) and that the arts (Galafassi et al. 72) and imagination (Milkoreit 4) can develop the type of thinking needed to create societal transformations. Teaching, reading, and discussing climate change literature and climate change arts can therefore be effective means of changing societal perceptions and generating climate change solutions.

Climate fiction is a form of speculative writing, and each story begins with a "What if. . . ." What if sea-level rise floods New York City? What if sand dunes form in the California desert, blocking escape from Los Angeles? What if Arizona or Texas becomes so hot as to be uninhabitable? These are among the "what ifs" present in *New York 2140*, by Kim Stanley Robinson; *Gold Fame Citrus*, by Claire Vaye Watkins; and *The Water Knife*, by Paolo Bacigalupi. The course I have titled Hot Mess, not because the course is a hot mess but because the planet is, begins with an analysis of "cli-fi" short stories in which I ask students to identify and analyze the "what if" presented in each story. Thinking about the "what if" helps students think in turn about the real-world climate change phenomena that are currently taking place. Seeing the scientific and sociological elements of climate change presented in a narrative and humanistic format, with characters, places, and real-world consequences, brings the science behind climate change to life.

As a next step, the class analyzes character tropes and symbols. An angry, middle-aged, white male character becomes a stand-in for climate change deniers in the United States. A scientist who has a poor publishing record and does not present practical applications for their research is a stand-in for, and a critique of, ineffectual scientific communication. A climate activist may be portrayed as being angry and equally ineffective at achieving their goal of social change. Those displaced by climate disasters represent the millions of real-world people displaced by the nexus of changing climate, extreme weather, and geopolitical turmoil. In *Flight Behavior*, by Barbara Kingsolver, the main characters include a scientist struggling to communicate science to the public, a reporter who is more concerned with reporting drama than with the facts at hand, and a lay-person caught in the middle, trying to understand the significance of the events occurring in her community. Each character becomes a stand-in for their representative segment of society: science, the media, and the public respectively. Analyzing these character tropes of climate change helps readers observe and question their operation in fiction and in life: Are activists always angry, pipe-bomb-wielding individuals? Can they be compassionate healers of collective environmental trauma instead? Must scientists be reclusive and locked away from the public they wish to bene-fit? Can they instead defy the status quo, take direct community action, and communicate their results in public forums?

Speculative fiction often emanates from a kernel of truth. The story seems plausible based upon what the author and audience know of the world right now. Since climate change is a scientific, social, and political phenomenon, my students and I take the time in class to identify these elements in the stories. We ask, What could be "ripped from the head-lines," as the saying goes? What plot point grew out of real-world science or potential real-world events? Drought and heat combine to create dev-astating fires, sea levels rise, species die off, governments partially collapse, and infrastructure breaks down—these fictional-narrative events can be matched to real climate-related events documented in the news from around the world or experienced firsthand in one's home state. Taken to-gether, these observations create a composite image of society experienc-ing climate change. They render visible what to some was once invisible.

Since 2008, the Yale Program on Climate Communication has admin-istered a nationwide survey on climate perceptions (Leiserowitz et al. 1). Beginning in 2015, I wanted to know my own students' perceptions of climate change and whether my teaching methods—namely, using cli-fi to educate students about climate change—were effective. I selected a sub-set of ten questions from the Yale survey and administered them to my students, first at the beginning of the course and again at the end. Re-sponses from eighty-eight participants, between 2015 and 2019, were re-corded using the Likert scale (Likert 11). Questions such as "How impor-tant is the issue of global warming to you personally?" showed that prior to the course, students were more concerned with global warming than the American public as a whole (Leiserowitz et al. 10), but their level of concern was not significantly changed after the course: students began and ended the course highly concerned. To a fundamental statement such as "Global warming is happening," eight respondents answered "neither agree or disagree" in the pre-survey, but following the course all students "agreed" with the statement. When questioned about the likelihood of global warming to harm people, students had different expectations about when and where this would happen. Prior to the course, twenty-four percent of respondents did not agree with, or were neutral about, the statement "Global warming is harming people now." After the course, this number was reduced to two percent, and there was a twenty-six percent increase in those who agreed with the statement a great deal. However, respondents showed no significant shift in perception of the statement "Global warming will harm people in industrialized countries." Before and after the course, they believed that global warming is unlikely to harm people in developed nations. In regard to "people in developing countries," there was a twenty-five percent increase following the course in those who said global warming will cause a great deal of harm. Possibly most important to those in the arts and humanities, there was a statistically significant change in perception for the statement "The arts have much to contribute to solving global warming." Perceptions shifted from "somewhat agree" to "strongly agree." In addition, answers to the question "How often do you discuss global warming with your family and friends?" showed a significant shift from "rarely" and "occasionally" to "occasionally" and "often."

Results from ecocritical controlled experiments and studies have shown that reader perceptions can be changed by reading climate fiction. Identification and transportation have been suggested as potential mechanisms for that change (Schneider-Mayerson et al. 9). Identification occurs when the reader connects to characters in the narrative. This is said to decrease social distance when the reader identifies with the locations and settings of a story (Scannell and Gifford 62). If the narrative tells the story of climate change, the reader can then begin to see themselves in the climate change scenario. Similarly, transportation is said to reduce the psychological distance between reader and character by focusing the reader's mental capacities on the narrative. In a climate change narrative, the reader becomes engrossed in the climate change scenes depicted. Reading cli-fi acts as mental simulation whereby the reader can begin to experience the happiness, sadness, fear, and threat from climate change encountered by the characters in the narrative (Milkoreit 4). Through transportation and identification in fictitious but plausible narratives, climate change becomes more real to the reader.

The way educators frame the discussion of climate change is important. Many young people today feel a paralyzing pressure to solve climate change. Educators can help by framing the discussion as one where the arts and humanities have a role to play in developing transformative solutions. The survey activity demonstrated that reading climate literature has the ability to change student perceptions. Specifically, those perceptions resulted in a greater appreciation for the arts as a change agent in solving climate change and a greater ability to discuss climate change outside of the classroom. Literature can develop the imagery, metaphors, and narratives of alternative futures needed for transformative social change. Climate literature can also help readers develop their imaginations to explore possible futures and creative solutions to climate change. If climate literature can change students' perceptions, it can change their minds. This is not only important in a society that is polarized around climate change, and where people are reluctant to accept science or change their views; it also suggests that the discussion on climate change can be reframed around the arts and humanities to increase student learning and real-world outcomes.

Works Cited

Bacigalupi, Paolo. *The Water Knife*. Orbit, 2015.

Corbett, Julia B., and Brett Clark. "The Arts and Humanities in Climate Change Engagement." *Oxford Research Encyclopedia of Climate Science*, edited by Matthew C. Nisbet, Oxford UP, 2017, pp. 1–27.

Galafassi, Diego, et al. "'Raising the Temperature': The Arts on a Warming Planet." *Current Opinion in Environmental Sustainability*, vol. 31, 2018, pp. 71–79. *Science Direct*, https://doi.org/10.1016/j.cosust.2017.12.010.

Kingsolver, Barbara. *Flight Behavior*. HarperCollins Publishers, 2012.

Leiserowitz, Anthony, et al. *Climate Change in the American Mind: December 2020*. Yale Program on Climate Change Communication, 10 Feb. 2021, pp. 1–90.

Likert, Rensis. "A Technique for the Measurement of Attitudes." *Archives of Psychology*, no. 140, 1932, pp. 1–55.

Milkoreit, Manyana. "Imaginary Politics: Climate Change and Making the Future." *Elementa: Science of the Anthropocene*, vol. 5, no. 62, 6 Nov. 2017, https://doi.org/10.1525/elementa.249.

Morton, Timothy. *Hyperobjects: Philosophy and Ecology after the End of the World*. U of Minnesota P, 2013.

Robinson, Kim Stanley. *New York 2140*. Orbit, 2017.

Scannell, Leila, and R. Gifford. "Personally Relevant Climate Change: The Role of Place Attachment and Local versus Global Message Framing in Engagement." *Environment and Behavior*, vol. 45, no. 1, 2013, pp. 60–85, https://doi.org/10.1177/0013916511421196.

Schneider-Mayerson, Matthew. "'Just As in the Book'? The Influence of Literature on Readers' Awareness of Climate Injustice and Perception of Climate Migrants." *ISLE: Interdisciplinary Studies in Literature and*

Environment, vol. 27, no. 2, spring 2020, pp. 337–64, https://doi.org/10
.1093/isle/isaa020.
Schneider-Mayerson, Matthew, et al. "Environmental Literature as Persuasion:
An Experimental Test of the Effects of Reading Climate Fiction." *Environ-mental Communication*, 15 Sept. 2020, pp. 35–50, https://doi.org/10
.1080/17524032.2020.1814377.
Watkins, Claire Vaye. *Gold Fame Citrus*. Riverhead Books, 2015.

Stephen Siperstein

Cli-Fi and Cultivating Cultural Agency

Students are spread out around the classroom, talking on their phones. No, we do not have lax classroom technology policies, and students aren't disrespecting our learning community. Rather, today in class I have specifically asked students to call the *FutureCoast* hotline and record voicemail messages from possibly climate-changed worlds ten to fifty years in the future. Some students are sitting clustered around tables in small groups, brainstorming and laughing together. Other students need a quieter space or are too bashful to "perform" in front of others and so step out into the hall. The room's atmosphere is lively, at times even cacophonous. The class's energy is high; we have recently finished reading and studying the more traditional literature of the course (short stories from the collection *I'm with the Bears* [Martin] and novels such as Nathaniel Rich's *Odds against Tomorrow* and Octavia E. Butler's *Parable of the Sower*) and are now shifting into an exploration of examples of multimedia climate change fiction (cli-fi) and students' creation of their own cultural texts. In this essay, I argue for the importance of including nonliterary cultural forms, particularly games, in the teaching of cli-fi (and in our broader understanding of climate change literature) and suggest that such projects provide fertile ground for students to exercise their own cultural

44

agency as cli-fi creators. Moreover, I suggest that the key function of cli-fi and climate change literature is their potential for cultivating community, both in the classroom and beyond.

Students in the class know that their voicemails will soon be published on the *FutureCoast* website for the world to listen to; one day, the recordings may even be preserved as "chronofacts" (physical objects on which voicemails are encoded) and geocached by some intrepid Futurecoaster (a *FutureCoast* role player). Today, my students are beginning to see themselves not just as readers or consumers of cli-fi but as scholars and planetary citizens (Siperstein et al. 14). They are authors, engaging in potent short-form storytelling about possible futures. They are part of a community of cli-fi creators, exploring "what if" scenarios in a collaborative narrative environment, and games like *FutureCoast* can enable this educational transformation.

In February 2014, the veteran game designer Ken Eklund teamed with the Polar Partnership at Columbia University to launch the *FutureCoast* project, an alternate reality game (ARG) that invited audiences to explore, create, and curate possible climate change futures and take part in collaborative storytelling. Eklund is the original designer of the *FutureCoast* game world, but he does not consider himself the author. Rather, he believes, "As a storyteller, as long as you keep it your story, your audience won't regard it as their story—and will be less affected by it" (Eklund and Thacher). By becoming cocreators in the narrative process, participants take more ownership over the direction of the story and are likewise more invested in the game's success. This is a key component of ARGs; as the game theorist Jane McGonigal explains, an ARG is "an interactive drama played out online and in real world spaces, taking place over several weeks or months, in which dozens, hundreds, thousands of players come together online, form collaborative social networks, and work together to solve a mystery or problem that would be absolutely impossible to solve alone."

FutureCoast built on the conceptual framework Eklund had established in 2007 with his award-winning alternate reality game *World without Oil*, which immersed participants in a fictional peak oil scenario and asked them to document their various experiences playing and living within this reality. In that game, participants not only imagined what it would be like to live through an oil crisis but also made choices in their day-to-day lives as if that crisis were happening and then shared their stories with others across a range of media. *FutureCoast* was similarly built on an imaginary premise: that a "leak" in the space-time continuum had caused

Figure 1. Geocaching a *FutureCoast* chronofact at a Eugene, Oregon, park. Photograph by the author.

voicemails from the future to appear in players' present reality. Game players could use their phones or the project's website to record a fictional voicemail from a climate-changed future. The game thus leveraged the power of the voicemail as a familiar cultural form to help participants grapple with the present and future impacts of climate change.

As a cultural genre with a set of shared patterns and expectations about sociality, the voicemail is both ephemeral and ubiquitous. Voicemails are intangible, fleeting packets of data and emotion. It is impossible to imagine holding a voicemail in our hands—but in *FutureCoast*, Eklund gave voicemails physical presence through objects he called "chronofacts." These small, futuristic-looking objects were sent to participants around the world to geocache in their communities during the six months in 2014 when the game was actively running (fig. 1). The participants would then broadcast across social media platforms the GPS locations of the chronofacts so that others could find and decode them by uploading discrete serial numbers onto the *FutureCoast* website. Doing this would "unlock" each associated voicemail for all players to hear. *FutureCoast* thus married a fanciful meta-narrative about a rupture in space-time with a quotidian cultural form.

Despite their ephemerality and ubiquity, or perhaps because of it, voicemails can be incredibly important and personal. We receive voicemails from family members and friends, doctors and colleagues, and strangers bringing unexpected news. Listening to some of the voicemails that my students recorded, which were later posted to the *FutureCoast* website, made clear to me just how intimate and emotional these microstories could be. For instance, a young daughter leaves a message for her mother during a powerful hurricane, saying through tears that this is probably the last time she will speak to her. In another voicemail, a young man says hello to his friend, then notices a "wall of water" moving rapidly and unexpectedly toward him. There's static followed by panicked yells and swooshes until the line eerily cuts out. Voicemails are rich with drama, information, and affect, often able to evoke in only one or two minutes an entire story, with plot, characters, and setting. In their reflections about the project, many students remarked on how much could be communicated just through the sound of a person's voice. One student wrote in their journal, "I loved this activity because it was personal and I could connect with some of the voicemails. One of the voicemails was someone calling their family to say goodbye because they don't think they will make it out of the disaster. It puts you in the shoes of someone else and you imagine if that will actually happen." By encouraging empathy and perspective-taking, the voicemails foregrounded the personal and emotional dimensions of climate change, engaging "both hearts and minds" (Eklund and Thacher). Another student similarly explained how the voicemail as medium helped her connect emotionally with the future: "[T]hese problems seem real because you can relate to the emotions that are in the voices of the people leaving these voicemails. There were a few voicemails that were so raw and emotional that I found myself caring for these people and wanting to help them. That is why *FutureCoast* is so powerful." Uploaded and collected on the *FutureCoast* website, the emotionally potent voicemails became part of a larger, anonymous collection of stories. They were no longer a private form, no longer locked onto a single cell phone, no longer "mine" or "yours." Instead, they became everybody's, an affective narrative archive of how people all over the world felt about climate change and the future.

In this respect, the game's curatorial function was just as important as the more active dimensions of game play, such as role-playing, recording voicemails, and searching for chronofacts. Visitors to the website, whether or not they recorded their own voicemails, could create playlists known as "timestreams" by selecting and organizing voicemails into categories such as "Extreme Events," "Technology Marches On," "Water, Water Everywhere," "Too Late," and "Before They Are All Gone . . ." (the last a collection of voicemails about extinct or endangered species). Once my students had created and recorded their voicemails, they were asked to create their own timestream. We discussed as a class the different ways in which one might choose to organize the voicemails (for example, according to tone, theme, or chronology) and to proceed from there into a larger conversation about cli-fi genre conventions, asking, What are the common themes and tropes in these voicemails? How are they similar to or different from the themes and tropes seen in cli-fi short stories and novels? What do such commonalities say about our collective approaches to imagining futures? Later in the term, I asked students to complete a similar task—creating their own cli-fi mixtapes—using all the texts we had encountered throughout the course, including the students' own works of cli-fi.

Drawing on McGonigal's notion of gaming for good, Eklund has described both *FutureCoast* and *World without Oil* as "authentic" and "serious" games. Authentic games, according to Eklund, are "multi-authored, textured in the way only diverse minds can supply; but also reality-based, painting reality within a playfully fictional frame" (Eklund and Thacher). Serious games are those "that intend more than entertainment for player[s]" and "aim to teach or train, often by realistically simulating some aspect of a world system" (Eklund). The outcomes Eklund envisioned for *World without Oil*, he has explained, relied on the relationship between systems thinking and affect: "In serious games, people get to play with complex systems that have a direct relationship to real-life systems. And these systems are complex and forbidding and yes, even scary, in the case of a global oil shortage. And so being able to experience them in an alternate reality helps us learn about them, and learning unravels fear" (Eklund). Research has shown that the experience of serious gaming can have significant impacts on players, including, for instance, heightened empathy and understanding of social and environmental injustices (Peng et al. 723; Pfirman et al.). Serious games have the potential to both enhance systems thinking and effect emotional or attitudinal transformations. As Eklund notes, "[S]erious games can be very effective at education, because they allow players to test and experiment with systems, which leads to better understanding of the relationships that comprise the system" (Eklund). *FutureCoast* required that its players experiment with the practices of speculation and with the ultimate complex system: the future. And because it was a game, albeit a serious one, *FutureCoast* was accessible and welcoming as it provided a platform for engaging with multiple perspectives concerning climate change. Eklund claims that one of his motivations in creating the game was to open up spaces for individuals from diverse populations to contribute their own personal ideas to collaborative visions of possible futures. "I think that there are a lot of people who want to have an invitation to say something about climate change," he has stated. "And I think this is the opportunity. It is this sort of creative challenge—you say it, but you say it in your future voice" (qtd. in Minchew). *FutureCoast* exploited the potential of cli-fi to open up how people think about climate change and invited them to talk about it:

> The challenge with climate change as a subject is the polarized state of its discussion. It's made people wary of engaging with anything that has the global warming or climate change label. Less well recognized, it's also disenfranchised people from the story—a scorched-earth war of talking points with no safe place left for the common person to venture hopes and fears or express what they know. (Eklund and Thacher)

FutureCoast invited people, regardless of background, ideological leaning, or worldview, to participate in imagining the story of climate change. It brought climate change into the commons as a topic of discussion and as a site for collaboration. Whereas much climate change discourse is global, abstract, polarized, and expert-driven, *FutureCoast* localized, personalized, depoliticized, and democratized the climate change conversation, thus breaking the "meta-silence" surrounding the issue (Marshall 82). Moreover, it turned the future into a site for play. As one reviewer of *World without Oil* remarked, "If you want to change the future, play with it first" (Olsen). Ultimately, then, the crucial component of *FutureCoast* was perhaps its most obvious: it was fun!

Because the experience of playing *FutureCoast* seemed to activate the students' creative energies, I dedicated the last weeks of the term to the final course project of creating cli-fi. As part of this assignment students conducted research about possible climate-changed futures, created their own work of cli-fi, situated that work within the genre more broadly by comparing it to other texts, close-read parts of their own work, presented their creations as part of a miniconference on the last day of class, and then reflected on their learning experience. Up until that point in the term, the coursework had emphasized analysis, the critical close reading and interpretation of literary and cultural texts, with few opportunities for students to exercise their own creative powers. This final assignment aimed to nurture the creativity students had first exercised while participating in *Future-Coast*. "Far too often," Elizabeth Tisdell writes, "we teach the importance of critique almost as if this is a form that transforms thinking, but we do not invite learners often enough to call upon the wonder of their own creativity. . . . This is about engaging more aspects of oneself that can lead to the transformation of *being* as well as thinking" (27). Creating cli-fi can indeed lead to the kind of ontological and emotional transformation that Tisdell describes here. Yet such transformation is not easy. Students who are used to being passive consumers of their education can find it especially difficult to shift to a model of learning in which they are responsible for designing something from scratch—in this case, creating a cultural artifact with few restraints on content or form.

Creating one's own work of cli-fi mere weeks after learning about climate change and encountering examples of the genre for the first time is a demanding undertaking. Though certainly meant to be fun, this creative project was time-consuming and serious, in Eklund's sense of the word. As someone who had been studying climate change intensively for over five years as well as, for most of that time, responding to my climate change experiences creatively (by writing lyric poetry), I did not fully grasp the magnitude of the challenge I had set my students. Therefore, I had to shift my expectations for the students' projects as well as recalibrate my own approaches to teaching. I discovered that I needed to model creative and critical speculative practice to provide students with a toolbox for engaging in this kind of work. That included taking on the role of coparticipant in establishing a learning community where students could take risks and play with new ideas and skills. It also involved acting as a coach, facilitating both self-directed and collaborative learning (including individual or group climate research conducted outside class) as well as ongoing metacognitive reflection about the learning process.

One of the biggest challenges that students faced in this project was constructing a convincing and immersive fictional space in which climate change presents itself as an immediate and pressing problem. Incorporating climate change into a work of fiction—whatever medium or form that work takes—is not a straightforward task, as students must consider how climate change affects landscapes both real and imagined, the political realm, economic systems, culture, human psychology, and so on. In their short story, "A New Dawn," one student drew on the conventions of cli-fi disaster narratives like *Odds against Tomorrow* and the film *The Day after Tomorrow* as well as information about the connections between climate change and the ongoing droughts on the West Coast. Narrated in the first person, "A New Dawn" tells the story of an individual who must grapple with the impacts of climate change; a fractured United States, now divided among the Confederated Christian Republic, the United Republic of North America, and the Cascadian Free State; and the discovery that bio-tech companies are genetically altering humans to manufacture energy through photosynthesis. The student conducted a significant amount of research, making use of both scientific sources and other literary works, and the story is thick with exposition about environmental politics and biotechnological developments. In their project reflection, the student admitted that it was difficult weaving exposition into the plot and ex-pressed a concern that "I felt I may have added too much information, and with that I fear I may have weakened [the story's] overall effectiveness as a piece of literature." Nevertheless, what the student saw as a concern I read as one of the story's strengths. Throughout the narrative, the pro-tagonist slowly uncovers information about his world, in both local and global terms, including insights into the connections between climate change and political corruption; this process models for readers how they too might engage in such discovery about their own world. Another stu-dent's cli-fi short story, "Survival of the Fittest," also incorporates ele-ments of disaster narratives but deploys them to explore a future in which the dual threats of climate change and gentrification force some residents of the city of Portland, Oregon, to become climate refugees. For this stu-dent, climate change is a pressing issue insofar as it acts as a threat multi-plier, exacerbating existing structures of social and environmental in-equalities.

As in these representative stories, most students chose to locate their cli-fi in places already familiar to them, and many reflected in their analy-ses that doing so helped them write with specificity and passion, regard-less of the form of the work. For example, one student's project, a video blog from the future, speculates on the multiple devastating climate change impacts on the Pacific Northwest region, thus functioning as a warning call for the present. Other students created games, audio drama podcasts, children's books, or storyboards for the novels they would have written had they had more time. Overall, instead of telling students what forms or media they should use, I asked them to explore, make choices, and re-flect on the successes and failures of those choices.

Many of the students' cli-fi projects and reflections attest to one of the crucial outcomes of this process: that students were able to exercise creative agency in making decisions not just about the form of their project but about the future itself. From their very first choices about story and plot, students faced innumerable questions about the relationship between climate change facts, their own imagination, and the future: Which set of predictions should I follow? Should I set my work in the near future, when changes might be more difficult to discern, or in a distant future, harder to predict, when changes might be drastic and more visible? Which threats are most serious or most likely? How do changing predictions about tipping points, extreme weather events, or rising sea levels affect my work's imaginative possibilities? Is it acceptable to oversell climate change threats or deviate from scientific predictions to provide more drama in the plot? Where should I set the work—in one location or many? Who are the characters, and how will they experience the material and psychological impacts of climate change? Will I include exposition, and if so, how much? Will I mention climate change explicitly, or will its impacts be visible in the setting, leaving readers to connect the dots themselves? Will I include potential solutions—political, economic, technological, cultural, or otherwise? Will the ending be hopeful or cynical, conclusive or open-ended? This list is not meant to be exhaustive, nor is it meant to suggest that every student considered all the questions. Rather, these questions foreground the transdisciplinary, synthetic, and imaginative dimensions of such work.

In her wide-ranging discussion of the engaged humanities, the Harvard scholar Doris Sommer suggests that "learning to think like an artist and an interpreter is basic training for our volatile times" (11). The process of creating cli-fi, whether by contributing fictional voicemails to a collaborative storytelling game, writing on a public course blog, or inventing cli-fi forms from scratch, opens a space for students not only to imagine possible futures but also to reappraise their own agency. Ultimately, this was a goal I only in retrospect discovered I had for my students and myself: to exercise what Sommer calls "cultural agency" (11) as a mode of grappling with the many challenges of climate change. According to Sommer, anyone, expert or not, can be a cultural agent, practicing the dialogue and coproduction of art that is at the heart of the engaged humanities.

Climate change education that emphasizes cultural agency and the production of creative artifacts (such as cli-fi) brings students together as authors and sharers, generating robust dialogue and collaboration in the classroom. The next step in developing the educational potential of cli-fi would be for students to collaborate with individuals and communities beyond the classroom. For instance, could the group of students who created a cli-fi children's book and a plan for using that book in elementary school classrooms have worked with teachers from a local school to bring the book in line with existing curricula and then gone into the classroom to pilot it? Or could they have helped elementary school students create their own cli-fi books? Could the student who created mock-ups for a cli-fi-themed set of large murals have worked with local community groups looking to incorporate more public art in the downtown area? Could the students who created mock-ups of different imaginative cli-fi games have collaborated with game designers outside academia to develop their ideas further? Or could they have brought their games to the many local game night events (at coffee shops and bookstores) around the city to work with actual game players? Yes, yes, yes, yes, and yes—if only we had had more than ten weeks.

In their treatise on the sustainable humanities, Stephanie LeMenager and Stephanie Foote argue that scholars and intellectuals in literary and cultural studies—including students—can take their work "further into the public sphere by acting more self-consciously as culture producers or allies of contemporary arts projects that reach communities outside the academy" (573). This requires reaffirming that work as "a kind of making" involving "the kind of collaboration that is established in other activist and scholarly communities" (574, 575). This practice of working with communities beyond the classroom can potentially have emotional benefits as well. Sommer refers to one such benefit as an "optimism of the will" that "drives life toward social commitments and creative contributions" (6). This, then, is one answer to that thorny yet ultimately galvanizing question: What can cli-fi, and climate change literature more broadly, do? "It won't do to indulge in romantic dreams about art remaking the world," Sommer explains, but neither "does it make sense to stop dreaming altogether and stay stuck in cynicism. Between frustrated fantasies and paralyzing despair, agency is a modest but relentless call to creative action, one small step at a time" (4).

In May 2014, Kate Schapira, a poet, climate activist, and writing instructor at Brown University, set up a booth in Burnside Park in Providence, Rhode Island, with a sign that advertised "Climate Anxiety Counseling, 5¢—The Doctor Is In." Though having no formal psychological or psychiatric training, Schapira wanted to get people talking intimately about a subject that troubled her greatly and hoped to "contribute to a shared language for talking about and responding to climate change and its effects" (Schapira, "Climate Anxiety Counseling"). Could a public park, she wondered, become an intimate setting where people could safely share their climate-change-related anxieties, fears, sorrows, and perhaps even hope? In part a whimsical tribute to Lucy's psychiatric booth from Charles Schulz's *Peanuts* comic strip, Schapira's project aimed to create a welcoming space for people to talk about their personal feelings concerning climate change. At first, as Schapira details on her website, not many people stopped to speak with her, and she received her fair share of odd looks from passersby ("Climate Anxiety Counseling"). Yet eventually people started stopping to talk (Schapira often waived the nickel fee). Thus began Schapira's impromptu climate change therapy sessions.

Schapira has since expanded her project to other locations, including parks and farmer's markets around Rhode Island, and asks "patients" to locate their worries on a large map of the state, creating a place-based visualization of the psychological impacts of climate change. With these "patients'" permission, she includes on the project blog stories about her experiences working as an informal climate change counselor. Schapira informs her interlocutors that she doesn't have any formal training in mental health care, and she emphasizes that "the booth is not set up for deep healing—if anything, it offers microhealings, sort of the opposite of microaggressions, things that are small on their own but that I hope have the potential to add up." But she speculates that it is precisely the informal character of the project that often draws people in and encourages them to open up: "In my little cardboard ramshackle booth, I don't look like I have a lot of power over other people—I don't look official—and I think for some people that might be what frees them to stop, and to speak" ("Points of Service").

Schapira conducts this work not only as listener, healer, mapper, and archivist but also as creator and author. In addition to writing about what happens and what she hears during these sessions, she extrapolates people's climate stories into what she calls "alternate histories," which could be considered a form of speculative cli-fi. Schapira starts with the climate anxieties that a "patient" shares and then composes a fictional story set in a future world in which that particular anxiety is no longer an issue. "By outlining some of what we could change or do differently in order to make that world possible," Schapira explains, the alternate histories provide a way to "reconsider what's necessary, what's habitual, what's structure and what's mutable about the world we live in now, and to help me, and hopefully you, imagine worlds that work better for more people, nonhuman creatures, and ecosystems" ("Alternate Histories: The Next Phase"). Thus Schapira makes the leap from hearing a personal story to imagining an alternative future. For example, in one story she imagines a Northeast region knitted together by high-speed rails as a response to the "patient" who worries about carbon dioxide emissions from cars and her own commute. In another, she imagines a world with no more industrial monoculture farms in response to a "patient" who worries about the effects of climate change on migrant agricultural workers ("Alternate Histories: 6/6, 4/23"). Schapira's "climate anxiety counseling" is thus not only a kind of grassroots mental health care but also a creative venture, a collaborative speculative project (perhaps one could call it counseling-booth-based cli-fi) that blurs the lines between imaginative art and radical social service.

I end this chapter with Schapira's project and this unexpected yet delightful scene of strangers talking intimately and personally about climate change in a city park as a provocation to think otherwise about how climate change storytelling and literature could function as an emotional and public resource in these difficult times. Schapira suggests that "we can take care of each other differently by turning some of the dials of expertise, intimacy, effort and protection to different levels" ("Points of Service"). More broadly, then, "climate anxiety counseling" exemplifies how climate change stories can help audiences encounter both possible futures and present realities, no matter how scary, together as a community of patients and practitioners, creators and caregivers, students and teachers.

Reading, writing, creating, and playing with cli-fi are practices not only for cutting through into possible worlds but also for reshaping the present. "My booth feels like action to me, though insufficient action," Schapira reflects. "But maybe [it's] a way to model the habits and interactions that can make our *present* more livable, more open, whatever it does for our future" ("Points of Service"). Ultimately, this is what climate literature does too, in the classroom and beyond.

Note

This essay draws from Siperstein, *Climate Change*.

Works Cited

Butler, Octavia E. *Parable of the Sower*. Four Walls Eight Windows, 1993.
Eklund, Ken. Interview by Dee Cook. *WorkBook Project*, 2 Mar. 2009. Accessed 3 Apr. 2016.

Eklund, Ken, and Sara Thacher. "Participatory Cli-Fi: The Making of Future-Coast." Interview by Stuart Candy. *Situation Lab*, 12 Feb. 2014, situationlab.org/futurecoast/.

LeMenager, Stephanie, and Stephanie Foote. "The Sustainable Humanities." *PMLA*, vol. 127, no. 3, 2012, pp. 572–78.

Marshall, George. *Don't Even Think about It: Why Our Brains Are Wired to Ignore Climate Change*. Bloomsbury, 2014.

Martin, Mark, editor. *I'm with the Bears*. Verso, 2011.

McGonigal, Jane. "Alternate Reality Gaming." MacArthur Foundation presentation, Nov. 2004, www.avantgame.com/McGonigal%20ARG%20MacArthur%20Foundation%20NOV%2004.pdf.

Minchew, Brandie. "The Future Is Fiction: Playful Future-Thinking about Climate Change with FutureCoast." *Alternate Reality Gaming Network*, 4 Feb. 2014, www.argn.com/2014/02/futurecoast_playful_future_thinking_climate_change/.

Olsen, Stefanie. "Provocative Politics in Virtual Games." *CNET*, 28 Mar. 2007, www.cnet.com/tech/gaming/provocative-politics-in-virtual-games-1/.

Peng, Wei, et al. "The Effects of a Serious Game on Role-Taking and Willingness to Help." *Journal of Communication*, vol. 60, no. 4, 2010, pp. 723–42.

Pfirman, Stephanie L., et al. "Engaging Systems Understanding through Games." *AGU Fall Meeting Abstracts*, vol. 1, 2013, ui.adsabs.harvard.edu/abs/2013AGUFMED13H..05P/abstract.

Rich, Nathaniel. *Odds against Tomorrow*. Farrar, Straus and Giroux, 2013.

Schapira, Kate. "Alternate Histories: 6/6, 4/23." *Climate Anxiety Counseling*, 4 Apr. 2015, climateanxietycounseling.wordpress.com/2015/04/23/alternate-histories-66-423/.

———. "Alternate Histories: The Next Phase." *Climate Anxiety Counseling*, 6 July 2015, climateanxietycounseling.wordpress.com/2015/07/06/alternate-histories-the-next-phase/.

———. "Climate Anxiety Counseling Is *In These Times*." *Climate Anxiety Counseling*, 10 Aug. 2015, climateanxietycounseling.wordpress.com/2015/08/10/climate-anxiety-counseling-is-in-these-times/.

———. "Points of Service: Responsive Art-Making and Intimate Public Discourse." *Climate Anxiety Counseling*, 19 Oct. 2015, climateanxietycounseling.wordpress.com/2015/10/19/points-of-service-responsive-art-making-intimate-public-discourse/.

Siperstein, Stephen. *Climate Change in Literature and Culture: Conversion, Speculation, Education*. 2016. U of Oregon, PhD dissertation.

Siperstein, Stephen, et al. Introduction. *Teaching Climate Change in the Humanities*, edited by Siperstein et al., Routledge, 2016, pp. 1–22.

Sommer, Doris. *The Work of Art in the World: Civic Agency and Public Humanities*. Duke UP, 2014.

Tisdell, Elizabeth J. "Themes and Variations of Transformational Learning: Interdisciplinary Perspectives on Forms that Transform." *The Handbook of Transformative Learning: Theory, Research, and Practice*, edited by Patricia Cranton and Edward W. Taylor, Jossey-Bass, 2012, pp. 21–36.

Jo Alyson Parker

Climate Change Stories:
Living and Dying in the Anthropocene

In his 2019 lecture on the temporality of climate change, the philosopher David Wood predicted a bleak future, but he also suggested that individual action might make a difference. Inspired by that message, I developed a new first-year seminar that I called Climate Change Stories. Such seminars, capped at eighteen students, are designed to introduce participants "to the adventures of learning in a college context, focusing in depth on a question or topic of disciplinary or interdisciplinary interest" ("General Education Program"). The format thus allows the students and me to approach climate change from diverse disciplinary perspectives and to consider how the issue affects all aspects of our lives.

I organize the course around three main units, each centered upon a key text supplemented by additional readings: "Future Shock," focusing on Octavia E. Butler's novel *Parable of the Sower*; "What Is the Sixth Extinction?," focusing on Elizabeth Kolbert's popular science text *The Sixth Extinction*; and "Living and Dying in the Anthropocene," focusing on David Mitchell's novel *The Bone Clocks*. Through a variety of fictional and nonfictional texts and assignments (reading responses, formal papers, videos, group work), students explore catastrophic climate change in order to understand what has brought humanity to this point, where we are now, what we might expect in the future, and what we can do in the light of this existential threat to humanity and to the planet.

Future Shock

I open this first unit with a rudimentary discussion of geologic timescales, including the contested term *Anthropocene*—the geologic time period defined by humanity's impact on the earth, encompassing geologic-scale changes to land, rivers, and seas and an alteration of the atmosphere. The first readings develop the idea of future shock. Elizabeth Kolbert's "Letter from Greenland" details the accelerating loss of Greenland's ice sheets and addresses the "time delay" that today's loss means for the future: "the warming that's being locked in today won't be fully felt until today's toddlers reach middle age. In effect, we are living in the climate of the past, but already we've determined the climate's future" (61). In "Learning to Die in the Anthropocene," Roy Scranton draws upon his experiences as a serviceman in a war-ravaged Iraq to extrapolate what the future holds for all of us because of catastrophic climate change, concluding, "The biggest problem we face is a philosophical one: understanding that this civilization is *already dead*." Students do group work on selected passages from the text and write a reading response to the following prompt: "How do Scranton and Kolbert show that these faraway places (Iraq and Greenland respectively) are harbingers of catastrophic climate change? What, if anything, in your own life already seems to be an indicator of catastrophic climate change?"

We then turn to *Parable of the Sower*. Written in 1993, the novel begins in 2024 and describes a dystopian United States ravaged by catastrophic climate change and the economic inequality that it exacerbates. The story is told from the perspective of Lauren Olamina, a young African American woman with "hyperempathy syndrome" (11), which causes her to feel what others feel, particularly pain. The text foregrounds the impact of climate change on the poor. Lauren's minimally subsisting California community has erected a wall to keep out those even poorer. Climate change has brought about persistent drought and drastic water shortages, destructive fires, and "early-season" hurricanes (15) that kill hundreds of people and destroy crops, leading to mass starvation. Lauren and her community contend with the elimination or reduction of social services (no public schools, pay-to-play fire and police departments). The plot details how Lauren—marginalized by race, sex, class, and ability—embarks on a quest to forge a new, self-sustaining community once her own is destroyed by marauders and fire.

During this unit, the students and I explore whether and how Butler's dystopian predictions of three decades past have already occurred. I bring in relevant videos, pictures, and charts: California's current drought readings, clips of the deadly wildfire that destroyed the town of Paradise in 2018, footage of fire tornadoes in Australia. Disturbingly, the Internet provides plenty of actual examples of Butler's futuristic imaginings. The first unit concludes with a formal paper assignment prompting students to consider how far along we are on the destructive path toward Butler's vision. I ask them to take a particular motif from the novel indicative of catastrophic climate change and the social breakdown that may occur as a result. Using newspapers, magazines, journals, newscasts, and documentaries to support their positions, students then explain how accurately Butler predicted the future or how she missed the mark. Students have addressed such motifs as increasing temperatures and drought, wildfires, walled communities, deadly storms, water scarcity, homelessness, increasing incidence of disease, and the growing divide between the rich and poor. Some have taken a creative approach, conceiving a present-day interview with Butler wherein she comments on how closely the situation today mimics what she described in her novel.

A doomed character in David Mitchell's *The Bone Clocks* warns that the future is "[c]oming soon to a Present near you" (58). Our first unit attempts to drive this message home, and students clearly show their grasp of it in their work.

What Is the Sixth Extinction?

I begin the second unit by showing *The Great Silence*, a video installation by Jennifer Allora and Guillermo Calzadilla, wherein shots of the Arecibo Telescope alternate with shots of endangered Puerto Rican parrots ("Allora and Calzadilla"). Overlaying the images is text provided by the science fiction writer Ted Chiang, purportedly from the perspective of one of the parrots. As the parrot poignantly tells us, "So the extinction of my species doesn't just mean the loss of a group of birds. It's also the disappearance of our language, our rituals, our traditions. It's the silencing of our voice" (Chiang 235). Students report that the video has a profound effect on them, and it provides a gripping lead-in to *The Sixth Extinction*.

In *The Sixth Extinction*, Kolbert explores the extinction of flora and fauna brought about by human action—and inaction. Each of the book's thirteen chapters focuses primarily upon a particular endangered or extinct species (such as the Panamanian gold frog or the great auk) or a destructive effect on an ecological system (such as the acidification of coral reefs or deforestation in Amazonia). The issues raised in each chapter may prompt readers to engage in further explorations.

Hence the second major assignment. Working in groups of three to four, students create an approximately seven-minute video from a written script deriving from one chapter in Kolbert's text. The assignment requires the students to sum up the key idea of the chapter, draw on research that goes beyond the chapter, and provide an annotated bibliography. Students come up with innovative ways of presenting their information. In spring 2020, for example, the group addressing Kolbert's chapter "The Sea around Us," on ocean acidification (*Sixth Extinction* 111–24), compiled a variety of video clips, charts, and graphs, and their voice-overs were lively and informative. The group dealing with "The Madness Gene," on the replacement of Neanderthals by modern humans (236–58), employed a "Draw My Life" format to explain what happened to Neanderthals, seamlessly meshing the drawings with informative voice-overs.

Living and Dying in the Anthropocene

We devote the remainder of the semester (typically six weeks) to a "slow time" reading of Mitchell's "Anthropocene novel" (Harris 152) *The Bone Clocks*, which spans a timeline stretching from seven millennia ago to thirty years into our future and addresses the human impact upon the planet. I assign Mitchell's novel because it makes clear how humans ignore the insidious, sometimes imperceptible, effects of climate change to our peril (see Parker).

The novel operates on two temporal levels. The first, spanning from 1984 to 2043, encompasses the lifetime of the protagonist, Holly Sykes—a lifetime during which the dire effects of climate change manifest. The second concerns a group of "atemporal" characters called Horologists (Mitchell 481), whose quasi-immortal lifetimes serve, as Rose Harris-Birtill has argued, "as a form of future-thinking temporality" (89) or "reincarnation time." According to Harris-Birtill, "What is important in such a system isn't a literal belief in reincarnation, but living *as if* the individual will be reborn to see their own behavioural consequences . . ." (105). Through the long-ago experiences of these fantastic beings, Mitchell addresses the centuries-old destructive impact of humanity upon the planet, and, through the idea that the Horologists will continue to be reborn and thus experience the consequences of their own actions, he prompts us, his readers, to be future-thinking regarding what *our* actions might entail.

In order to emphasize this future-thinking quality of Mitchell's text, I draw students' attention to two relevant websites. For her Future Library initiative (www.futurelibrary.no), the artist Katie Paterson asks renowned writers to contribute a manuscript that will remain unread for one hundred years before it is published; a recently planted forest in Norway will supply the paper upon which the books will be printed. Mitchell, the second writer to contribute a manuscript, has said of the project, "For me, it's a vote of confidence in the future of culture. The project is a declaration of belief that, one century from now, despite the threats to civilization posed by climate change and its deniers, by racist demagogues and by death-cultists, our great-grandchildren will still value trees, books, reading and narrative" (qtd. in Mackin). The other initiative, the Long Now Foundation, is intended "to foster long-term thinking and responsibility in the framework of the next 10,000 years" ("About Long Now"). To that end, the foundation is building a clock designed to keep time for ten thousand years. As the builders ask, "If a Clock can keep going for ten millennia, shouldn't we make sure our civilization does as well?" ("Clock"). Like Mitchell's novel, both initiatives prompt us to think beyond our own limited lifespans—to be future-oriented—and to consider whether such future-oriented thinking would have helped or can help mitigate the effects of climate change. (Indeed, we may wish to consider whether a future-thinking in terms of just a few months might have mitigated some of the catastrophic effects of the COVID-19 pandemic.)

I cannot discuss Climate Change Stories without addressing the impact that the pandemic had upon the 2020 version of the course and upon my students' thinking about the future. Although by the time this essay is in print, this particular pandemic may have been declared over, the lessons it has taught and is teaching us with regard to climate change will persist, including the ideas that humanity's destructive impact on wildlife habitats may have triggered the disease (see, for example, Bruillard). It will take many cohorts of students before those lessons are forgotten.

In 2020, we did not return to the university after spring break. The climate crisis, whose slowly accreting, catastrophic effects are on a time delay that will catch up with us around mid-century, was eclipsed by another crisis whose world-shattering effects took place, it seemed, almost instantaneously. Both crises, however, twine together, and my new goal was to help students draw connections between what they were facing as a result of the pandemic and what they would face in a couple of decades if the drivers of climate change went unchecked.

Although dealing with climate change, *The Bone Clocks* provided an eerily prescient view of the social dysfunction catalyzed by the pandemic. During the eighteen-year gap that occurs between its penultimate section, "A Horologist's Labyrinth," and its final section, "Sheep's Head," the novel's characters enter the "Endarkenment"—a time of deprivation, fear, and, as the name implies, ignorance, superstition, and brutality. The stark contrast between the earlier sections and "Sheep's Head" drives home how quickly social order can disintegrate. Granted, the eighteen years that separate the last two sections of *The Bone Clocks* represent a much longer time span than our world's plunge in just a matter of weeks into our own variety of the Endarkenment. Situations and events in Mitchell's text, however, parallel situations and events that have occurred in our pandemic-inflected society. The question that generated the most responses on our discussion board asked students to consider what one aspect of life in "Sheep's Head" reminded them of pandemic-era life today. Students noted the lack of resources (tampons and insulin in Holly's society, toilet paper and PPE in ours), struggling economies, pervasive uncertainty, and anxiety in both societies.

As initially conceived, the final paper assignment asked students either to envision what sort of manuscript they would write for Future Library, and what kind of readership would be around to read the book, or to imagine what kind of life they would be living in 2043. When I began to prepare my course for online delivery, I offered students a third option: to compare and contrast Holly's 2043 situation with the situation in which they found themselves living now. The strongest papers played riffs on these options as the students imaginatively braided together descriptions of their lives in the present, their visions of the future, and their understanding of the course themes that we had addressed.

One student, Drew Noto, proposed writing a Future Library manuscript, geared toward people his age, that would detail his day-to-day life as a new college student dealing with a pandemic. Drew noted that, as COVID-19 upended his life, he wished that he could find a comparable text exploring the 1918 pandemic from the perspective of an eighteen-year-old. In envisioning a future readership, he astutely remarked that, just as the last great pandemic had occurred one hundred years in the past, another pandemic might take place one hundred years in the future, and its impact would be compounded by the ravaging effects of catastrophic climate change. His text, he hoped, would show that people had rallied together and made it through the crisis, thus bringing hope to a future readership.

Jenna Reyes wrote letters to "Future Jenna," detailing increasing environmental destruction. Younger Jenna pointed out how the situation actually worsened postpandemic: "The increase from gasoline-powered vehicles, coupled with the gases released from the re-opened factories, puts us two steps back to where we were before the pandemic. . . . People nowadays are too concerned with the stability of the economy rather than the stability of the Earth." Future Jenna writes in turn to her daughter, offering a poignant apology for a ravaged earth: "I'm sorry that my generation failed you. . . . The coronavirus acted as a catalyst to the industry which took so many years off the Earth's lifespan. I don't know how long we have until everything is gone. I'm sorry we left you with a broken home."

Lucia Nardelli contrasted her prepandemic life as a carefree student with her pandemic-era stay-at-home life, explaining the devastating sense of loss she felt. She then predicted what sort of societal changes might occur once some semblance of normalcy returned and imagined how this time in crisis might be assessed in the future: "My generation will be talking about the COVID-19 pandemic, just as the generations above me talk about challenging periods of time such as the [1918 pandemic] or the Great Depression. I think this pandemic will be a warning for future generations that one person's actions can affect a whole world"—as we see with our society's heedlessness regarding climate change. Referring to Mitchell's novel, Lucia nevertheless ended her piece on a high note:

> Reading about the Endarkenment in the story is quite eye-opening because it is a possibility of a future for our world, which makes me want to have a positive impact on society. . . . I hope this hard time our society has been facing will open doors to positive changes for our world. I have been motivated to make the best of every day offered by my hopes for what the future is to hold once this crisis comes to an end.

Although behind pandemic-induced anxiety is the even greater anxiety induced by catastrophic climate change, teaching Climate Change Stories—and teaching *The Bone Clocks* during a pandemic—has taught me that my students' resilience, creativity, and determination may help ensure that the world of "Sheep's Head" does not come to pass.

Note

The idea of a "slow time reading" comes from Paul Harris, who teaches Mitchell's text in his first-year seminar on the Anthropocene at Loyola Marymount University. I would like to thank Harris and Ann Green for suggestions in designing the course, Tom Weissert for reading an early draft of the paper, and Chontel Delaney and Karen Pinto from Saint Joseph's University Academic Technology and Distributed Learning for providing helpful guidance to the students as they made their videos.

Works Cited

"About Long Now." *Long Now*, longnow.org/about/. Accessed 25 Mar. 2023.

"Allora and Calzadilla (in collaboration with Ted Chiang), *The Great Silence*." *Vimeo*, uploaded by Artribune Tv, 14 Dec. 2016, vimeo.com/195588827.

Bruillard, Karen. "The Next Pandemic Is Already Coming, Unless Humans Change the Way They Interact with Wildlife, Scientists Say." *The Washington Post*, 3 Apr. 2020, www.washingtonpost.com/science/2020/04/03/coronavirus-wildlife-environment/.

Butler, Octavia E. *Parable of the Sower*. 1993. Grand Central Publishing, 2019.

Chiang, Ted. "The Great Silence." *Exhalation: Stories*, by Chiang, Alfred A. Knopf, 2019, pp. 231–36.

"The Clock of the Long Now." *Long Now*, longnow.org/clock/. Accessed 17 July 2023.

"General Education Program: Signature Core Requirements." Saint Joseph's University, sites.sju.edu/geprog/gep-requirements/signature-core-requirements/.

Harris, Paul. "David Mitchell's Fractal Imagination: *The Bone Clocks*." *Substance: A Review of Theory and Literary Criticism*, vol. 44, no. 1, 2015, pp. 148–53.

Harris-Birtill, Rose. *David Mitchell's Post-Secular World: Buddhism, Belief, and the Urgency of Compassion*. Bloomsbury Academic, 2019.

Kolbert, Elizabeth. "Letter from Greenland: A Song of Ice." *The New Yorker*, 24 Oct. 2016, pp. 50–61.

———. *The Sixth Extinction: An Unnatural History*. Picador, 2014.

Mackin, Laurence. "Writers Blocked: The New Works by David Mitchell and Margaret Atwood That You Can't Read till 2114." *The Irish Times*, 25 May 2016, www.irishtimes.com/culture/books/writers-blocked-the-new-works-by-margaret-atwood-and-david-mitchell-that-you-can-t-read-till-2114-1.2660512.

Mitchell, David. *The Bone Clocks*. Random House, 2014.

Parker, Jo Alyson. "Mind the Gap(s): Holly Sykes's Life, the 'Invisible' War, and the History of the Future in *The Bone Clocks*." *C21 Literature: Journal of Twenty-First-Century Writings*, vol. 6, no. 3, 2018, pp. 1–21, https://doi.org/10.16995/C21.47.

Scranton, Roy. "Learning to Die in the Anthropocene." *The New York Times*,

10 Nov. 2013, archive.nytimes.com/opinionator.blogs.nytimes.com/2013/11/10/learning-how-to-die-in-the-anthropocene/.

Wood, David. "Founder's Lecture: Is Time Out of Joint? Or at a New Threshold? Reflections on the Temporality of Climate Change." *Time in Variance*, edited by Arkadiusz Misztal et al., Brill, 2021, pp. 43–65. The Study of Time 17.

Sofia Ahlberg

The Anthropocene as a Global Coming-of-Age Story: A Pedagogy in Transition

Can literature save the planet? I worked this question into a short survey when introducing teacher trainees to the English education program at Uppsala University, Sweden. My survey contained a mixture of lighthearted and serious questions that received a range of responses. Surprisingly for me, when I put the question "Would you love to convince students that reading can save the planet?" to forty teacher trainees, ninety percent of them responded, "Yes." Although the teacher training program offers a range of reading from Christina Rossetti to Claudia Rankine, it is the coming-of-age story, with its vivid depictions of personal growth, that typically elicits the most interest. This is not so surprising since this genre reflects students' own struggles toward maturity in a rapidly changing world and informs their training to become teachers who will one day support others' coming-of-age. That the school curriculum inadequately meets the demands of young people growing up in times of climate crisis has been made abundantly clear by movements such as Fridays for Future. As the young climate activist Alexandria Villaseñor says of the purpose of Earth Uprising, a nonprofit organization that focuses on peer-to-peer climate education, "Until you grown-ups get it together to improve school curricula, young people can teach one another what's happening to our planet and how to help mitigate the climate crisis" (326). The basis of my approach to teaching literature in times of crisis is what I see as a teacher's imperative to help students cope with anxieties by helping them develop confidence in their own abilities as well as trust in others sufficient for them to engage in collaborative activities. At the same time, affirming the Anthropocene as a species-wide coming-of-age for us humans rather than an ending helps me frame lessons that can instill hope rather than despair.

One way to improve school curricula suggested by members of Earth Uprising is to give literature teachers in training the tools to apply critical reading skills to real-world events. In my own classes I focus especially on teaching the relevance of the coming-of-age story within the young adult genre, since research suggests this genre has the potential to support readers in making "difficult life decisions that may have been previously unconsidered or viewed as impossible" (Alsup 213). The popularity of recent coming-of-age trilogies over the last decade, ranging from the dystopian *The Hunger Games*, by Suzanne Collins, to the more recent fantasy *Legacy of Orisha*, by Tomi Adeyemi, suggests that students are likely to have already engaged with the coming-of-age theme set against a diverse range of backgrounds. This prior knowledge can be used to scaffold learning in the literature classroom. Through the exploration of narratives of self emerging within social upheaval, students may find reading that resonates with their own experiences as well as learn more about intention, action and consequence, and conflict and resolution. Making literary studies keenly relevant for students requires that we underscore that they can learn how the world works through reading.

I believe that as teachers it is our duty to impart grounds for measured hope in the face of increasingly dire predictions for climate change. Teachers need to find a balance between a stance of cautious hope—seeing the climate crisis as an opportunity for change, for example—and one that responsibly acknowledges the seriousness of climate-related events. What is required is a pedagogy in transition, one that does not succumb to the gloom-and-doom scenario but rather sees the classroom as a future-oriented environment where students can imagine and practice the roles they will play in facing a post-Anthropocene future. While coming-of-age stories are generally about the maturation of an individual, the most memorable are those in which central characters discover in themselves the capacity to discern social problems that have gone unnoticed as well as the expressive power to enlist the support of others in projects to redress these ills. As a rule, these stories see the protagonist contesting existing hierarchies of power and privilege. In what follows, however, I read the Anthropocene itself as a coming-of-age story. There are two reasons for this: First, what we know so far about the present human-dominated geological epoch known as the Anthropocene obliges the heaviest users of energy to grow up and realize that a high-octane lifestyle has very serious, life-threatening environmental consequences. Second, by interpreting the Anthropocene as a coming-of-age story rather than a story of inevitable decline, teachers can nevertheless inspire hope by focusing on the significance of species-wide growth spurts and transitions and not just individual development.

Even though scientists have been considering the environmental impacts of human life since the nineteenth century, it was the visible presence of a global anthropogenic signature preserved in the geological record from around the mid–twentieth century that gave rise to the concept of the Anthropocene (Turney et al.). Some count the beginning of agriculture eight thousand years ago as the start of the Anthropocene; others cite the increase in human dependence on fossil fuels halfway through the last century, which fueled a growth spurt also known as the Great Acceleration—a development that, despite perceived benefits, greatly magnified the environmental impact of industrialization. At the core of the idea of a human-dominated geological period is the harmful concept of progress as growth and the aspiration to climb the ladder of economic success. That aspiration is mirrored in the economic rise of the West from about 1610 and the eventual establishment of global trade links. These phenomena depended and still depend on colonial exploitation and enable an unjust sense of modernity whose unfair trade practices created the "Third World" as "backward, irrational, poor," and so forth (Andreotti 103). The Anthropocene is thus also the mark of an "uneven global imaginary and its by-products (nationalism, exceptionalism, consumerism, materialism and individualism) as we enjoy the (false) sense of stability, fulfillment and satisfaction that they provide (belonging, community, togetherness, prestige, heroism and pride)" (103–04). To address the climate crisis, it is crucial that we imagine ways of coming of age that go beyond these historical constructs. We need to challenge these conventional views of the modern self by developing alternative models of growing up in the Anthropocene that can help redefine it as a transition period rather than an ending.

Traditionally, the bildungsroman genre stems from an Enlightenment conception of *Bildung* (education, development, and progress) that often overlooks the sometimes violent and traumatic social development of marginal or less privileged subjects (Castle 369). This has given way to narratives that reflect systemic injustice and oppression as these conditions impact identity, education, nationality, and social relationships. The contemporary bildungsroman is openly apocalyptic, often with an explicitly ecological lesson addressed to a young adult readership (Matz 270). Indeed, the young adult dystopian novel often presents environmental disaster as the catalyst for its hero's maturation (Basu 7). Much of the appeal of the dystopian coming-of-age novel lies in the celebration of an individual leader who succeeds in replacing an unjust or corrupt establishment with a new order more representative of the values of the younger generation. Although this is a rich field for research, I concur with Greg Garrard, who argues that apocalyptic rhetoric at times "fosters a delusive search for culprits and causes" (115). For this reason, in this essay I avoid a focus on the explicitly dystopian and instead bring into discussion aspects of the coming-of-age story that put pressure on what Western readers conventionally mean by societal and individual progress.

In Paolo Bacigalupi's short story "Pocketful of Dharma," state-sanctioned deprivation of the poor and extreme centralization of power are imaged in the form of an enormous living skyscraper named "Huojianzhu," a hybrid organic edifice towering over Chengdu in central Sichuan, China. At the beginning of the story, the young beggar boy Wang Jun yearns to be admitted to this living structure of power purpose-built to supply the rich with affordances no longer available to ordinary people in a ravaged world. What makes Wang Jun different from other beggars is that he possesses a data cube especially sought after by the ruling elite because it contains the consciousness of the Dalai Lama. By the time we reach the end of the story, Bacigalupi's hero does in fact attain "the luxury of the heights of which he had always dreamed," but with that comes the realization that he no longer aspires to be part of the territorializing design of Huojianzhu. Having become embroiled in a geopolitical struggle over possession of the data cube, Wang Jun prefers to descend from the heights, taking his cube with him, "slipping deeper into the mist, clambering for the slick safety of the pavement far below." He chooses to remain incognito down below and not use his access to the Dalai Lama to advance his situation.

This story presents Wang Jun's coming-of-age as a riposte to the world of power and influence. Rather than enter into a struggle with the ruling elite, he refuses to engage with them. In this way, the story offers an alternative to what Nancy Lesko and Frank Topping see as a rather monolithic construction of coming of age based on historical assumptions about what constitutes progress: "Adolescence became a social space in which progress or degeneration was visualized, embodied, measured, and affirmed. In this way, adolescence was a technology of 'civilization' and progress and of white, male, bourgeois supremacy" (29). Countering such limited assumptions about what constitutes progress—whether personal or social—is key to finding ways to address or negotiate climate change or other crises that have their roots in these assumptions. I find discussion questions such as the following useful for discerning in the coming-of-age story its potential to recalibrate thought and action and to challenge ingrained beliefs about what constitutes growth and well-being:

> What forms of consent or dissent are displayed in the story?
> What forms of conformity or resistance are displayed?
> What is the protagonist's role in addressing conflict?
> In what ways does the protagonist enact or reject the role of hero?

Reading the Anthropocene as a species-wide coming-of-age is a way of staging a pedagogy that can extract important and hopeful lessons from gestures like Wang Jun's refusal to parlay his spiritual stake for personal advancement. At the same time, this method also allows readers to see in coming-of-age stories examples of heroism that can guide them in the Anthropocene. This period demands a new kind of maturity in humans that refuses predetermined narratives of progress that perpetuate social injustice. Of significance here is an emphasis on fostering and maintaining social and natural relationships, which is why I value collaborative and explorative reading practices over solitary reading during times of crisis. What gives Wang Jun the confidence to choose the descent over an aspiration to power and fame is his realization that a private moment of edification, however humbling, is more valuable than a position as a pawn in a game not of his own design.

After the discussion prompted by the questions listed above, students are invited to complete what I call the storyboard exercise. There are two parts to this exercise. First, I ask students to squeeze the whole of the story or novel under discussion into just five key scenes or storyboards, as if they were preparing to adapt the text to the screen. I do this to encourage students to synthesize by thinking about what contributes the most to a character's growth as well as what precisely is meant by "growth" to start with. Each student's five storyboards will also differ greatly from the others and so bring to the surface a variety of reasons for choosing particular scenes. I'm inspired here by Hilary Janks's approach to teaching literacy, which aims to empower students with the insight that access to opportunities or "life chances" is dependent on social position. These components of her model—"power, diversity, access" (320)—contribute to the fourth component, which is "design/redesign," meaning that whatever is critiqued or deconstructed at the level of social reality can be redesigned through language. Janks provides the following elaboration: "once you've deconstructed a practice or a text or a behavior, or anything that's taken for granted, what do you do about it? And that's the redesign process that leads to social action" (321). To begin to address that question, in the second part of the storyboard exercise, I invite students to substitute the five "scenes" they settled for with material from their own lives, as if they were adapting the text to a more familiar setting where their own input is easier to imagine. The only requirement is that they still focus on the themes already touched on in the previous discussion.

I have chosen Bacigalupi's story because of its intrinsic interest as well as its contemporary take on coming of age. But it should be clear that it is not essential to address the method introduced here only to the teaching of coming-of-age stories. It is possible to include other narratives that may not strictly be considered coming-of-age but may inform discussion of a maturing process that can affect society as a whole or even our entire species. Thus, to name a well-known canonical text, I would happily teach *The Great Gatsby* as a story in which Nick Carraway is still developing, along with America, a properly mature assessment of the limitations of the American dream. It would be clear that that myth is one that can no longer be sustained through the Anthropocene. By acknowledging the growing importance of collaboration and protest in our age, as well as critiques of traditional forms of individualistic agency, this approach provides a framework for discussing points of contemporary relevance even in texts from earlier periods and other genres. What is important is that we seek out narratives that can support a teaching of the Anthropocene hopefully, as a stage humanity will pass through, and realistically, as an epoch with painful lessons for our species to learn.

Works Cited

Alsup, Janet. "Female Reader Reading YAL: Understanding Norman Holland's Identity Themes Thirty Years Later." *Young Adult Literature and Adolescent Identity across Cultures and Classrooms: Contexts for the Literary Lives of Teens*, edited by Alsup, Routledge, 2010, pp. 205–15.

Andreotti, Vanessa. "The Educational Challenges of Imagining the World Differently." *Canadian Journal of Development Studies / Revue canadienne d'études du développement*, vol. 37, no. 1, 2016, pp. 101–12.

Bacigalupi, Paolo. "Pocketful of Dharma." [1999]. *Baen*, www.baen.com/Chapters/1597801348/1597801348___1.htm.

Basu, Balaka, et al. Introduction. *Contemporary Dystopian Fiction for Young Adults: Brave New Teenagers*, edited by Basu et al., Routledge, 2013, pp. 1–15.

Castle, Gregory. "Coming of Age in the Age of Empire: Joyce's Modernist Bildungsroman." *James Joyce Quarterly*, vol. 50, nos. 1–2, 2012, pp. 359–84.

Garrard, Greg. *Ecocriticism.* 2nd ed., Routledge, 2012.

Janks, Hilary. *Literacy and Power.* Routledge, 2010.

Lesko, Nancy, and Frank Topping. *Act Your Age: A Cultural Construction of Adolescence.* Routledge, 2012.

Matz, Frauke. "Alternative Worlds—Alternative Texts: Teaching (Young Adult) Dystopian Novels." *Learning with Literature in the EFL Classroom*, edited by Eisenmann W. Delanoy and Matz, Peter Lang, 2015, pp. 263–80.

Turney, Chris S. M., et al. "Global Peak in Atmospheric Radiocarbon Provides a Potential Definition for the Onset of the Anthropocene Epoch in 1965." *Scientific Reports*, 19 Feb. 2018, www.nature.com/articles/s41598-018-20970-5.

Villaseñor, Alexandria. "A Letter to Adults." *All We Can Save: Truth, Courage, and Solutions for the Climate Crisis*, edited by Ayana Elizabeth Johnson and Katharine K. Wilkinson, One World, 2020, pp. 323–27.

Matt Burkhart

Apprehending Climate Change through Fiction and Film

In *Slow Violence and the Environmentalism of the Poor*, Rob Nixon highlights the word *apprehension* as a critical term that "draws together the domains of perception, emotion, and action," underscoring that to apprehend is "to perceive" and "to arrest" (14). Further, Nixon identifies the ways that "states of apprehension—trepidations and forebodings" characterize affective responses to global environmental concerns such as climate change (15). Over the years I've been teaching Cli-Fi: Apprehending Climate Change in Fact, Fiction, and Film, Nixon's layered meanings of *apprehension* have cohered as concepts governing the course. Since the spring of 2017, I have taught ten sections of this course, which is a writing-intensive seminar tailored to sophomores and juniors at Case Western Reserve University. The seminar serves core requirements and draws students who are already engaged in climate mitigation efforts, such as the campus hub of the Sunrise Movement and the Student Sustainability Council, as well as others who enroll in the course strictly to fulfill graduation requirements.

At the semester's outset, I use in-class writing assignments to take stock of perceptions of climate change, prompting students to address how they apprehend the changing climate in their daily lives and how they think it affects places, beings, or practices they care about. In classes dominated by students pursuing degrees in engineering and physical sciences, establishing the group's baseline awareness creates a platform for discussing the transporting, imaginative qualities of literature. During the initial weeks of the course, I work with students to assemble a critical tool kit, which includes excerpts from Nixon's text alongside a range of scholarly and journalistic texts. While manifestations of climate change become more evident with each passing year, we discuss how the world-building capacities of literature and film allow us to apprehend those phenomena at the scale of the global, the national, the regional, and the local. Namely, we discuss how narratives transport audiences across space and time, allowing readers and viewers to envision how the livelihoods of spatially distant, contemporary people in coastal or arid environs are already profoundly impacted by climate change.[1]

During the first week of class, we read sections from Elizabeth Kolbert's *Field Notes from a Catastrophe*, attending to her use of long-form journalism deeply informed by literary tropes. We consider how Kolbert's depictions invite readers to imagine the daily lives of Inupiaq facing the relocation of their longstanding home in Shishmaref and of field scientists on the Alaskan Northern Slope and the Arctic Ocean whose research sites are disrupted by thawing ice. Kolbert establishes the attunement to close observation of biota through the seasons that allows field scientists and people who have lived close to the land for many generations to offer credible insight into increasingly dramatic change over time. Inspired by class discussions during the first semester I taught Kolbert's book, I have come to frame her text as characterizing climate change as a creeping, haunting specter that gradually insinuates itself into the daily lives of people, starting at places that readers in the continental United States might regard as the earth's far-flung margins. Specifically, Kolbert establishes how centuries-old, Indigenous hunting and fishing practices have changed in just a decade (7–8), how real estate purchases in Alaska are impacted by sinkholes caused by melting permafrost (15), and even how such a sinkhole opened up and split a house in two (16). I amplify students' initial observations about how climate change is undermining people's capacity to develop a stable sense of home, ranging from disruptions of familiar subsistence practices and threatened relocation from islands to be subsumed by rising seas to domestic structures that are literally being torn asunder. I suggest that Kolbert's telling can be illuminated through the lens of gothic haunted house stories, like Edgar Allan Poe's "The Fall of the House of Usher." Readers may recall how that story's narrator highlights the correspondence between an increasingly animistic setting—a widening crack that runs from the mansion's roof down its facade, extending to a nearby glacial tarn—and taboo transgressions against domestic propriety committed by the surviving scions of a fabulously wealthy family. I use the occasion to address the centrality of concepts like "the uncanny" and "the return of the repressed" (Freud, *Uncanny* and "Repression" 156) to this text—and to gothic fiction in general. Specifically, I encourage students to think how Freud's etymology of the uncanny excavates its rootedness in the *unheimlich*—or the rendering inhospitable of what was once "homely" (*Uncanny* 1). We discuss how Kolbert's writing about senses of home fractured by climate change suggestively aligns with the climax of Poe's story, in which a widening crack cleaves the manor house when a great tempest pummels the land under preternatural moonlight. Ultimately, the tarn's edges are breached, and its water rushes through the widened chasm to subsume the home's ruins.

I have weighed whether overlaying that gothic narrative onto Kolbert's writing is warranted but have felt affirmed that it frontloads for students some ideas that we later encounter in excerpts from Amitav Ghosh's *The Great Derangement*. Ghosh foregrounds the vitalism observed in the land in cultures deeply rooted in India's Sundarbans and contrasts it with daily life in the Anglosphere, whose conveniences are made possible through the parlaying of colonial accumulations of land and wealth into carbon-intensive infrastructures (5, 10). As a class, we discuss implications of Ghosh's contention that anglophone authors of the past two centuries have focused on human interiority in ways that occlude the ecological costs of transforming the land to suit settler colonial economies and ideologies (7, 80).

In the semester's early weeks, we also catalog Kolbert's methods of artfully conveying scientific detail to her readers. Specifically, I ask each student to pick out and share a concept that is new to them or a rhetorical flourish that struck them as memorably compelling—or perhaps a bit "off." These low-stakes initial discussions allow students to hear from all their classmates, to share impressions from the reading, or to dig in more deeply according to their capacity as critical readers. Those conversations frequently make note of Kolbert's painstaking efforts to frame technical terminology, including "positive feedback" (31), "albedo" (29–31), "thermohaline circulation" (56), and "thermokarsts"(15), in a colloquial fashion that helps readers apprehend their significance both at the micro level of lived human experience and the macro level of processes unfolding around the globe.

We juxtapose our reading of Kolbert's haunting account with excerpts from David Wallace-Wells's *The Uninhabitable Earth*. Specifically, we discuss how Wallace-Wells constructs a persona that appeals differently to readers than Kolbert's self-presentation. By establishing himself as someone who does not consider himself a conventional environmentalist, thus as someone who can bring fresh eyes to climate data, Wallace-Wells clears out space to adopt a more strident and alarming tone that students find convincing. From this place, he underscores his urgent concern for the future his daughter and later generations will inherit. Whereas Kolbert is heavy on thick description and attention to how climate change will broadly impact biotic communities, Wallace-Wells offers an approach that on balance is more anthropocentric and favors the mode of the jeremiad, which he uses as a vessel for numerical predictions and scientific modeling that communicates the macro-impacts of our looming failure to reach goals that limit greenhouse gas emissions.

In an early three-page writing assignment, I ask students to make a case for the relative merits of these texts by Kolbert and Wallace-Wells as well as Jeff Orlowski's documentary film *Chasing Ice*. While that triad of texts hardly offers an exhaustive range of narrative approaches, it allows students to consider the impact of tone and persona as well as the communicative possibilities presented by written media and visual media respectively. Leading up to that brief paper, we observe that *Chasing Ice*, like Kolbert's writing, hinges on efforts to translate scientific data—such as rates of glacial melting, temperature increase, and information found in ice core samples—into meaningful frames of reference that render the magnitude of change apprehensible to viewers. We also note the dramatic treatment of the knee injuries that threaten to prevent the documentary's central figure, James Balog, from continuing his glacial photography. By that point in the semester, we have also read Ghosh's critique of the modern Euro-American fascination with "individual moral adventure" (76–81, 127–31) as privatizing our consciousness but also atrophying our capacity to track stories focalized through many perspectives across space and time (76–77). Nevertheless, when discussing the melodrama with which Balog is portrayed as soldiering through pain to capture images of disappearing glaciers, I am careful to not overplay this critique of "individual moral adventure." I have found that Balog's distinctive account speaks particularly to student athletes, who describe identifying with the calculated risk that goes into putting one's body on the line for a larger collective athletic endeavor. These students' insights offer humbling reminders that different types of appeals can resonate deeply with some people's affective experience. Such moments remind me of the merits of pragmatically structured syllabi and discussions, which may focus on texts I find less compelling but may also offer some students a meaningful point of entry into overarching climate change conversations.

That first four-week unit culminates in a longer paper wherein students are tasked with making a case for how some combination of approaches from our initial critical and nonfiction readings could help humanity avert a future like the one featured in Naomi Oreskes and Erik M. Conway's work of speculative history, *The Collapse of Western Civilization.* Through that assignment, I establish a context in which students can think about the third facet of Nixon's use of *apprehension*—namely, how we might summon tools of perception, affective reflection, representation, and political efficacy in hopes of arresting or at least mitigating the interlocking positive feedback loops that exacerbate the climate crises.

As the class transitions into addressing novels and feature films, we spend the next seven weeks discussing how authors and filmmakers toggle the variables of time in ways that allow us to imagine the future consequences of past and present choices. By that, I mean that we mark out how authors and filmmakers manage audiences' relationships to temporality in ways that shape the agency of characters and readers or viewers in texts set in four general time frames: the recognizable present (Kingsolver's *Flight Behavior*; Rich's *Odds against Tomorrow*; *Beasts of the Southern Wild*), the near future (Butler's *Parable of the Sower*; Bacigalupi's *The Water Knife*), more remote postapocalyptic futures (*Mad Max: Fury Road*; *Snowpiercer*; Robinson's *New York 2140*), and alternation between the accessible past and the near future (Boyle's *A Friend of the Earth*). During any given semester, my syllabus includes three novels, each paired with two or three documentary or feature films. Students are asked to watch films outside class as their primary assignment before the next class period. In response to each film, at least one student shares their research findings through a contextualizing presentation that frames our class discussion of how the film and the novel at hand complement each other and provoke new lines of discussion.

Constant on my syllabus throughout all iterations of the course has been Octavia E. Butler's *Parable of the Sower*, which I accompany with viewings of *Mad Max: Fury Road* and *Beasts of the Southern Wild* to highlight these works' implication of petroculture and of patriarchal narratives that inure audiences to socially exaggerated scarcity in ways that perpetuate oppressive hierarchies. As with *Parable*, both films invite identification with female protagonists who creatively navigate climate-changed worlds rather than hewing terminally to "business-as-usual" structures. Each semester has also featured Paolo Bacigalupi's *The Water Knife*, paired with Jennifer Baichwal and Edward Burtynsky's documentary collaboration, *Watermark*, and Bong Joon-Ho's *Snowpiercer*. While *Watermark* offers stylized amplification of ideas central to Marc Reisner's *Cadillac Desert*, which serves as a MacGuffin in *The Water Knife*, the train in *Snowpiercer* offers an icy analogue to Bacigalupi's Sonoran Desert dramatization of "arcologies" as profoundly exclusive, green-adapted living structures for the economic elite. (Bacigalupi borrows the portmanteau *arcology*, fusing *architecture* and *ecology*, from Paolo Soleri, who used the term to characterize his efforts to develop sustainable, high-density communities in resource-strained locales like the desert Southwest [see "Arcology Concept"]. While Soleri's vision was egalitarian in nature, Bacigalupi depicts arcologies as exclusionary enclaves where only the most affluent can properly shelter themselves from the inhospitable conditions that less privileged people must endure.) Both *Snowpiercer* and *Water Knife* foreground climate refugees' experience in ways that allow students to discuss ethical tensions central to the novel, which Bacigalupi frames in terms of seeing the world through "old eyes" or "new eyes." Similar concerns invite comparative discussion of how the two texts manifest Christian Parenti's term "the politics of the armed life boat" (qtd. in Ghosh 143), referring to Malthusian efforts to insulate wealthy individuals, corporate entities, and nations that historically have produced outsized greenhouse gases from the displaced people who feel climate change's impacts most heavily (Ghosh 143–49).

In addition to considering the ethical implications of speculative fiction, we grapple with the intertwining of artifice and social context, examining authors' use of recognizable social phenomena to dramatize the role of social inequity in the radically transformed biosphere of the climate-altered futures we seek to avoid. To spur discussion of the novels' aesthetic innovations and cultural contexts, I assign contextualizing presentations and point students toward initial resources, such as central critical, historical, and literary texts they can consult to draw links between research findings and textual dynamics. Presenters are responsible for modeling the use of contextual approaches to frame interpretations of the primary text. They also facilitate discussion of at least two questions they have developed to share with their classmates. In this role, presenters mediate between the primary text and cultural contexts, often illuminating points of connection between texts that I have overlooked.

For instance, through contextualizing presentations, students help their classmates understand how Butler's *Parable of the Sower* dramatizes the ways that multinational corporations supplant civic structures by presaging the impacts of trade agreements like NAFTA, including the looming threat of debt peonage and neoslavery in company towns. For instance, students presenting on that topic have integrated materials from historians and cultural geographers that help their classmates understand similarities between company towns (from earlier moments in the carbon-fueled economic expansion of the United States), the living conditions surrounding maquiladoras proliferating in the early years following the passage of NAFTA, and the conditions of profound unfreedom that underlie the promise of economic and personal security in Olivar, a dystopian gated city run by a trinational corporation, which Butler calls Kagimoto, Stamm, Frampton and Company. In Butler's novel, all these circumstances are exacerbated by rising global temperatures that result in drought and meteorological catastrophes. Butler dramatizes the deleterious impacts of petroculture, in which fossil fuels are used solely by multinationals' transport trucks and by bands of marauders addicted to Pyro, a drug that imparts ecstatic pleasure to accelerating the world's burning through the use of fossil fuels. This transpires under the governance of President Donner, aptly named to evoke the party of westward settler colonists whose navigational misfortune led them to resort to cannibalism. We discuss how Donner is cannibalizing opportunity for present and future generations by imposing neoliberal austerity measures and how the novel's protagonist, Lauren Olamina, fiercely advocates for a far-reaching embrace of intergenerational concerns that unites communities around an ethics of care and compassion for the most vulnerable. To similar ends, student presentations on Bacigalupi's *The Water Knife* illuminate the foregrounding of multinational water compacts and the dehumanization of refugees in ways that point to recognizable present conditions: to the inequitable relations that govern the US-Mexico border, to the plight of contemporary climate refugees, to the flux of migration spurred by dust bowl conditions earlier in US history, and to a climate-change-driven trajectory that will reduce the United States to the conditions of a failed state if we don't forge more equitable relations at the level of environmental, economic, and social justice.

Whereas the two novels just discussed highlight protracted drought as the dominant expression of climate change, Nathaniel Rich's *Odds against Tomorrow* features a Hurricane Sandy–like event that catastrophically inundates New York City. Its protagonist and narrator, Mitchell

Zukor, is a recent University of Chicago graduate and a member of Generation Seattle (witness to the decimation of Seattle by earthquake) who is morbidly obsessed with disaster. His dire projections serve, homeopathically, to keep his memories at bay and to inoculate his consciousness against deeply feared future disasters. In his work as a consultant for Future World, Mitchell indemnifies corporate clients against liability for disaster—not through any actual mitigation of potential threats but through a legal loophole allowing payment of princely sums for disaster consulting in lieu of action. Among other contextualizing approaches, one of the student presentations for this novel helps us understand how *Odds* is indebted to Don DeLillo's *White Noise* in its use of dark humor to grapple with environmental disaster. Following the critical trajectory established in Joseph Meeker's *Comedy of Survival*, I help students appreciate how both novels exemplify the tragicomic mode in contemporary environmental literature. By that, I mean that Rich and DeLillo traffic in dark humor as the specter of mortality looms heavily over both texts. However, rather than offering lugubrious laments over the death of nature in the tragic funereal mode, both hint at an uneasy marriage between humans and the other-than-human world. For DeLillo, this takes the form of nightly reverie at the sublime sunsets made possible by an "airborne toxic event"; in turn, near the conclusion of Rich's novel, Mitchell has spearheaded the reinhabitation of Brooklyn flatlands near Jamaica Bay that have been scoured by the hurricane's storm surge. He contemplates his dominion over this urban homestead, but just as he is about to take his axe to a fallen oak, signs of luminous decay catch his eye and he is entranced by a "grotesque insectopolis" of "crawling, munching, slurping, rotting, liquifying, cannibalizing" (291). Contrasted with Robinson Jeffers's musings in an earlier century on his potential "enskyment" were he to be eaten by a vulture, Mitchell humors the notion of being devoured by this egalitarian empire of the abject. When contextualizing the novel's conclusion through the lens of William Cronon's critique of the retreat to wilderness, we explore how the novel's denouement suggests an Anthropocene-epoch back-to-the-land movement, where survivors of climate catastrophes can direct efforts toward the regeneration of high-density urban locales with the same half-aware reverence with which an earlier generation fled the industrial and sociological complexities of cities in the 1960s and '70s. The resettlers of Brooklyn are less marked by vulnerable racial and class identities than the climate-displaced persons alongside whom Mitchell is temporarily housed in a Randall's Island (NY) FEMA camp. In our discussion of this whiter, back-to-the-land-styled resettlement of Brooklyn, we take into account interpretations such as Ashley Dawson's and Peter Moskowitz's, which frame the post-Katrina diaspora of Black residents from hurricane-affected wards in New Orleans as a mode of climate-leveraged gentrification through which displaced people of color effectively face multiple barriers to returning and rerooting in communities they have called home for generations (Dawson 233–35; Moskowitz 16–20, 24–26, 45–49, 65–67).

Slightly past the semester's midpoint, I assign a four-to-five-page essay that tasks students with putting at least two fictional works in conversation with at least one of the critical or contextual approaches encountered early in the semester or through class presentations. Like a Melvillean tryworks, the essay should serve as a forum for students to synthesize and distill their thinking, perhaps into a foundation for their research-intensive final paper. For that final essay, students are required to make a persuasive case regarding the role of fiction (novels or dramatic films) in the public understanding of climate change and asked to craft informed speculation regarding how audiences might respond to at least two works of climate change fiction. As our seminar's culminating class activity, we devote several class periods to delivering "works in progress" presentations. I give students eight to ten minutes to present their thesis, their critical approach, and what they see as their best pieces of citable evidence from their research findings and their primary texts. I also challenge them to pull back the curtain on the ways their projects feel unfinished. Rather than ending on a note of decisive conclusion, I encourage students to affirm how they see their best evidence and reasoning communicating effectively with their intended audience but to also explicitly indicate the aporia in their writing process—that is, where they have hit walls in their research, where they feel they are pushing the outer limits of a theoretical concept's applicability, or where they are less confident that their audience would find their approaches convincing. We then take about ten minutes to respond to each presentation, working collectively to remind one another of concepts that might offer leverage or to share research findings that could work toward mitigating the vexing problem of climate change. For each presenter, I assign two primary respondents who are responsible for initiating in-class commentary and can start online conversations with the presenter through *Google Docs*. This functions like a dynamic and ongoing peer evaluation that continues through the writing process, ensuring that each student has at least two classmates they can turn to who can offer familiar and engaged commentary on their paper's development.

When setting up these presentations, I convey that there is a premium on integrating novel outside research through our campus library as well as from a broad swath of popular sources, yet I also frame our multiday works-in-progress conference as a forum for generously exchanging ideas. Given the magnitude of the climate crisis, alongside expanded awareness of how students' desires to mitigate climate change can be too easily framed as an issue resolvable through green consumerism and other minor modifications of a business-as-usual system, I urge students to move beyond atomized, individualistic approaches: to remind one another of earlier readings and to point their classmates toward sources they might find helpful. That is, in the same pragmatic spirit in which I structure my syllabus to offer some students a meaningful point of entry into overarching climate change conversations, I encourage students to move beyond business-as-usual approaches as they forge a greater synthesis that incorporates not just superficial tweaks but some fundamental reconsideration of solutions driven by consumer capitalism. In hopes of a general recalibration of values on a campus where students frequently strive to competitively distinguish their engineering prowess or their candidacy for medical school, I encourage participants in my seminar to envision themselves as generous researchers both by keeping an eye out for references that might support their classmates' research and by sustaining ongoing dialogue that supports their peers' revision process.

As in the vehicle of metaphor, my abiding goal is to figuratively transport students through this course to future or imagined worlds that, together, we wish to avoid in order to discern ethical paths that might help us collectively disrupt trajectories that will render the earth's biosphere uninhabitable for humans and the many species with which we have coevolved. In so doing, I prepare students to apprehend the epiphenomena of climate change, and they reckon with affective dimensions of apprehensiveness and grief that arise with a transforming world. This serves the goal of working to identify the beliefs, values, and practices that have led to our current juncture and the third register of Nixon's invocation of *apprehension*—that is, in hope of arresting those trajectories and initiating paths that generate meaning, value, and energy that might sustain humanity and the biosphere for generations to come.

Note

1. We take as an example statistics regarding the preponderance of North Americans who don't dispute that climate change is happening and is anthropogenic yet still envision it as so distant in space and time that it will not manifest as an exigent concern for them in their lifetime ("What"). Speaking to the capacity of literature to drag that temporal horizon into more immediate view, we also discuss Paolo Bacigalupi's pointed comment that speculative fiction "allows readers to connect with their future selves" ("Paolo Bacigalupi").

Works Cited

"The Arcology Concept." *Arcosanti*, 2023, www.arcosanti.org/arcology/.

Bacigalupi, Paolo. *The Water Knife*. Alfred A. Knopf, 2015.

Beasts of the Southern Wild. Directed by Benh Zeitlin, Fox Searchlight Pictures, 2012.

Boyle, T. Coraghessan. *A Friend of the Earth*. Penguin Books, 2001.

Butler, Octavia E. *Parable of the Sower*. Warner Books, 2000.

Chasing Ice. Directed by Jeff Orlowski, Exposure Labs, 2012.

Cronon, William. "The Trouble with Wilderness; or, Getting Back to the Wrong Nature." *Uncommon Ground: Rethinking the Human Place in Nature*, edited by Cronon, W. W. Norton, 1995, pp. 69–90.

Dawson, Ashley. *Extreme Cities: The Peril and Promise of Urban Life in the Age of Climate Change*. Verso Books, 2017.

DeLillo, Don. *White Noise*. Viking Penguin, 1985.

Freud, Sigmund. "Repression." 1915. *The Standard Edition of the Complete Psychological Works of Sigmund Freud*, James Strachey, translator and general editor, vol. 14, pp. 141–58, Hogarth Press / Institute of Psycho-analysis, 1955.

———. *The Uncanny*. 1919. Translated by David McClintock, Penguin Books, 2003.

Ghosh, Amitav. *The Great Derangement: Climate Change and the Unthinkable*. U of Chicago P, 2016.

Jeffers, Robinson. "Vulture." *The Selected Poetry of Robinson Jeffers*. Edited by Tim Hunt, Stanford UP, 2001, p. 697.

Kingsolver, Barbara. *Flight Behavior*. HarperCollins Publishers, 2012.

Kolbert, Elizabeth. *Field Notes from a Catastrophe*. Bloomsbury, 2015.

Mad Max: Fury Road. Directed by George Miller, Village Roadshow Pictures, 2016.

Meeker, Joseph W. *The Comedy of Survival: Studies in Literary Survival*. Charles Scribner's Sons, 1974.

Moskowitz, Peter. *How to Kill a City: Gentrification, Inequality, and the Fight for the Neighborhood*. Nation, 2017.

Nixon, Rob. *Slow Violence and the Environmentalism of the Poor*. 2011. Paperback ed., Harvard UP, 2013.

Oreskes, Naomi, and Erik M. Conway. *The Collapse of Western Civilization: A View from the Future*. Columbia UP, 2014.

"Paolo Bacigalupi: What If the Drought Never Ends?" *YouTube*, uploaded by FORA.tv, 4 Aug. 2015, www.youtube.com/watch?v=AZLEDsddI34.

Poe, Edgar Allan. "The Fall of the House of Usher." *Tales of the Grotesque and Arabesque*, Philadelphia, 1840, pp. 75–104.

Reisner, Marc. *Cadillac Desert: The American West and Its Disappearing Water.* Viking Penguin, 1986.

Rich, Nathaniel. *Odds against Tomorrow.* Picador, 2014.

Robinson, Kim Stanley. *New York 2140.* Orbit, 2017.

Snowpiercer. Directed by Bong Joon-Ho, Opus Pictures, 2013.

Wallace-Wells, David. *The Uninhabitable Earth: Life after Warming.* Tim Duggan Books, 2019.

Watermark. Directed by Jennifer Baichwal and Edward Burtynsky, Sixth Wave Productions, 2012.

"What Climate Change and Coronavirus Have in Common." *Vimeo*, uploaded by Department of Political Science, 3 Feb. 2021, vimeo.com/507933531.

Part II

Locations

Christina Gerhardt

Sea-Level Rise, Low-Lying Islands, and Caribbean Literature

Caribbean island nations are among the nations that have contributed the least to carbon dioxide emissions and global warming; they are, however, suffering their effects already, severely and disproportionately. Climate-change-induced sea-level rise and more frequent and intense hurricanes are already impacting archipelagos in the Caribbean. When teaching literature from the region, I begin with spatial and temporal reconfigurations.

Given that the students taking the course and the educator teaching it will vary in terms of demographics, life history (where one has lived, among which communities, etc.), and area of study or expertise, I try to gather us to consider space. Many area studies tend to focus on continents, so I work to make students aware of the importance of decontinentalizing the gaze, as it were. Here, one could assign Brian Russell Roberts and Michelle Ann Stephens's introduction to *Archipelagic American Studies.* Albeit focused on American studies, its lessons hold true for other area studies within modern languages (anglophone, francophone, etc.). Or, depending on the course's level (graduate or undergraduate), its home base (American studies, Caribbean studies, English, etc.), and its length (quarter or semester), one could read it as an educator and share the map on the cover and inside of Daniel Immerwahr's *How to Hide an Empire: A History of the Greater United States.*

Then I shift the focus onto the archipelagic, onto islands and the relations among islands. Here, again depending on the level of the students, we might read essays by the Martinican Édouard Glissant in the collection *Poetics of Relation.* Countering the fellow Martinican Aimé Césaire's concept of négritude as too focused on the relationship to Africa, Glissant proposes the notion of *antillanté,* rooting Caribbean identity within the Caribbean. He also argues for an archipelagic framework for rethinking relations among islands.[1] Glissant focused not only on the Antilles but also, via *Créolité,* on the Creole Caribbean—that is, not a monolithic but a diffuse, diverse, complex, and heterogeneous Caribbean—in history, language, and culture, including Indigenous Caribbeans, Black Caribbeans, East Indian Caribbeans, and more. Doing so undoes the colonial reconfiguration of the region, which has rent islands asunder via different colonial structures, histories, and languages.

We rethink not only space but also time. Given that the course engages climate change, we consider the politics of dating of the Anthropocene. Various dates and rationales have been put forward in texts that are foundational, with regard to the relevant environmental science and the environmental humanities and also to courses within this field or that engage it. These works include Dipesh Chakrabarty's "The Climate of History"; chapter 6 of Donna Haraway's *Staying with the Trouble: Making Kin in the Chthulucene* (99–103); Simon Lewis and Mark Maslin's "Defining the Anthropocene"; Jason Moore's introduction to *Anthropocene or Capitalocene? Nature, History and the Crisis of Capitalism*; chapter 2 of Kathryn Yusoff's *A Billion Black Anthropocenes or None* (23–64); and Heather Davis and Zoe Todd's "On the Importance of a Date; or, Decolonizing the Anthropocene." Discussing these texts, we consider the implications of dating and the varying structures of the arguments made. These questions are not only temporal but also spatial. For example, does it make sense to date the Anthropocene to the advent of the steam engine in the United Kingdom? Or is it, as Kathryn Yusoff, citing the Jamaican author Sylvia Wynter's "1492: A New World View," argues, better to start in 1441, when Portuguese colonizers began enslaving and abducting Africans from Senegal and western Africa (Yusoff 14; see also Spillers)?

Starting on the west coast of Africa is not only a temporal and spatial reconfiguration and does not only return the Black labor (as important as the Indigenous land[2]) to the narrative of colonization; it marks the moment, as Wynter writes in "On How We Mistook the Map for the Territory," when what constitutes the human (as well as humanism and the humanities) began to be structured by the other—where race and racism were established.[3] These considerations provide historical, economic, environmental, cultural, and philosophical framings for a discussion of the region's literature that engages climate change.

To segue from the historical-political-economic frameworks of colonialism to present-day imperialism and the impacts of the climate change in literature, we read and discuss Audre Lorde's "Grenada Revisited: An Interim Report" and Dionne Brand's "Nothing of Egypt," both about the 1983 US invasion of Grenada. Lorde's essay considers the United States' history of invasion and ongoing occupation of the previously independent islands Puerto Rico, Guam, Hawai'i, and Diego Garcia and the use of Vieques, Guam, Kaho'olawe, and Diego Garcia for US military exercises, showing how this invasion took place on the heels of and in the context of the US fear of the then burgeoning self-liberation and self-determination or independence movements in Southeast Asia, Africa, and the Caribbean. Brand's essay discusses how her experience of the US invasion affected her. This essay supplies critical historical context not only of colonization but also of present-day imperialism that informs strategies for addressing climate change. These issues include historical debt (owed to and not by Haiti), reparations, and historical inequities arising from colonialism, which are now compounded by climate change. While the Global North disproportionately has emitted and is emitting carbon dioxide, the impacts of the climate crisis have been and are more intensely felt in the Global South, which is financially less able to address these effects. Against this background, we discuss Grenada's 1974 independence and the hurricanes that struck that nation in 1979, 1980, 2004, and 2005.

By now we are halfway through the semester (and halfway through the essay). Why allocate so much time to these historical considerations and frameworks? I think it is important that students spend a lot of time discussing colonialism and related frameworks before I ask them to consider climate change, its impacts, possible solutions to it, and the literature that engages it. I hope that such classroom discussions help students understand the importance of not repeating structures that have unleashed inequities on already historically traumatized populations.

Geologic matter is a critical predictor of the impacts of sea-level rise. Islands that are low-lying and consist of limestone will be more intensely affected by sea-level rise and hurricanes. Limestone is the Swiss cheese of geology. It is the most permeable and porous geologic matter. It allows salt water to intrude. Some even say it soaks up salt water—like a sponge. As a result, when sea levels rise, limestone islands will be inundated not only at the shoreline (from sea-level rise and storm surge) and from above (stalling hurricanes and their rains) but also from below, as water can percolate up through the porous material. Just as vital are the geologic layers of history, or history as geologic layers. It is key to be attentive to the different layers and to read them together, in conversation, as Glissant suggested via his concept of *Créolité*. History exists in multiple registers. It is polyphonic, not conforming to a single voice, as Katherine McKittrick has said (McKittrick et al.). It is a chorus, as Christina Sharpe has put it ("In the Wake"), not only in terms of people but also in terms of reading people and matter together and of reading with a temporal span, a history and a future in mind. With ears attuned and eyes refocused on how the layers shape one another as one listens anew and looks anew, we continue on our journey.

If these historical events lay a geologic foundation for the present, how do they manifest in literature? Texts by writers such as M. NourbeSe Philip and Brand provide different answers. In the note after *Zong!*, an antinarrative that transforms legal text into poetry, Philip writes, "There is no telling this story; it must be told." *Zong!* tells of the eponymous ship that "in 1781 . . . leaves from the West Coast of Africa with a cargo of 470 slaves and sets sail for Jamaica. . . . Instead of the customary six to nine weeks, this fateful trip will take some four months on account of navigational errors of the captain . . ." (189). Claiming that the food and water on board were insufficient for survival, and believing that the trip's insurers would reimburse the ship's owners for the financial loss, the captain ordered 130 enslaved Africans thrown overboard. When the ship owners claimed compensation, they were refused and brought a case against the insurers. "The report of that decision, *Gregson v. Gilbert*, the formal name of the case more colloquially known as the *Zong* case," Philip, a former lawyer, writes, "is the text I rely on to create the poems of *Zong!* To not tell the story that must be told" (189). Katherine McKittrick, engaging *Zong!* (and more), writes:

> This is a relation story nested in black studies, in science studies of blackness and race, and black creative text. Divided into different sections, the story uses columns, repetition, narrative blocks and academic form to center black ways of knowing. The method . . . will observe and critique the injustices of racism without revering and repeating and describing racial violence. . . . This is a story that lives with violence but

cannot accurately describe violence. This is a story where I live with violence without knowing how to full comprehend, or detail, violence. (*Dear Science* 125)

Philip describes her process as follows:

My intent is to use the text of the legal decision as a word store; to lock myself into this particular and peculiar discursive language in the belief that the story of these African men, women, and children, thrown overboard in an attempt to collect insurance monies, the story that can only be told by not telling, is locked in this text. In the many silences within the Silence of the text. I would lock myself in this text in the same way men, women, and children were locked in the holds of the slave ship *Zong*. But this is a story that can only be told by not telling, and how am I to not tell the story has to be told. (191)

Philip ruminates on silence and on the stories that can and must be told: on their absence in an obvious format, on their presence, on access to them, and on their requiring a different approach or method of listening and writing—one that brings them to the surface but also avoids reinscribing the violence within which they are embedded. *Zong!* required different narrative strategies in its writing and requires different methodologies in its reading.

Narrative rupture manifests too in Brand's *A Map to the Door of No Return*, where the author writes, without initially mentioning it by title, about her novel *At the Full and Change of the Moon*. "This novel," Brand writes, "begins in a museum," referring to a fort on Trinidad, Fort King George, "laid down around 1783" (196). She moves through the museum's rooms and floors. "Fear of disrespect to something quite old makes me linger but then sheepishly move on. . . . This novel begins as I move to the staircase to the second floor. . . . Those bones warn me that everything after I have made up, I have invented in absence" (198–99). She reads the writing of Thomas Jefferys, geographer to the king, George III, who writes in a strangely elegant prose. "This gorgeous prose dissembles, it obstructs our view of its real directions, it alludes, it masks. But it points, it says, there, that is where you land the ships bringing slaves to this island. It says that it is possible to do this and still maintain gifts of erudition, or intelligence, even playfulness. Language is wonderful, so deceitful" (202). Care can be brought to history—or not. "Which is why," she continues,

230 years later I wrench it from his pen, I tear it from the wall of this museum, I cut it into pieces—one piece for the title of this novel, *At*

> *the Full and Change of the Moon*, and the rest I give to my Kamena . . .
> who is searching in this novel for a place he will never find. . . . He
> never finds what he is looking and longing for, it eludes him, it dis-
> sembles, all of his directions lead him nowhere. (202)

Brand writes of futile search, of dispersal, of scatterings: "I cannot unhap-
pen history and neither can my characters." She continues, citing Derek
Walcott, "When asked, as in Derek Walcott's poem, 'Where are your mon-
uments, your battles, martyrs? / Where your tribal memory?' my characters answer as in that poem, 'Sirs, / in that grey vault. The sea. The sea/
has locked them up.'" Brand writes that her "characters can only tear into
pieces, both history and Jefferys' observations" (203). It is about the story
told and the story not told, about the dissembling and the scattering.

Reading *Zong!* and Brand's *A Map to the Door of No Return* or *At the
Full and Change of the Moon*, students become attuned to a different way
of writing and of reading. They became attuned to how absences can be
written and heard. Much of *Zong!* is about engaging absence, silence, written over (as Brand describes it above). Philip writes, "I witness a continuation of my engagement with the idea of Silence vis-à-vis silence begun in
Looking for Livingstone: There I explored it as one would a land, becoming aware that Silence was its own language that one could read, interpret, and even speak" (195).[4] Philip looks and listens for silence, lingers
with it, and returns voices to the record without speaking over or for.
Christina Sharpe, in *In the Wake*, writes about "all of the circulating images of and in the aftermath of Hurricane Katrina and the 2010 earthquake in Haiti" and how "such repetitions often work to solidify and make
continuous the colonial project of violence. . . . What kinds of ethical viewing and reading practices must we employ. . . . What might practices of
Black annotation and Black redaction offer?" (117). The redaction means
the withholding or withheld or unstated, or the stated that is spoken for.
The annotation means looking for signs of the life, of what McKittrick
calls "Black livingness." Rather than tending toward Black death, McKittrick encourages scholars look for "Black life" and "Black livingness."[5]

These texts work differently on us. They demand that we do different
types of work. When we turn in the class, then, to Caribbean literature
that grapples with the impacts of climate change mentioned at the outset,
such as sea-level rise and hurricanes, having discussed the theoretical stakes
of dating the Anthropocene, with their temporal and spatial implications,
and literature that provides a deeper context for reading the present mo-

ment, students engage it more substantively. They are looking more deeply for what a text does and does not say, does and does not include. We linger, considering what might constitute an ethical viewing or reading and what it might reveal.

We conclude with a reading of Alexis Pauline Gumbs's *Undrowned: Black Feminist Lessons from Marine Mammals*. Gumbs, the author of an earlier trilogy (*Spill*, *M Archive*, and *Dub*) engaging the work of the Black feminist theorists Hortense Spillers, M. Jacqui Alexander, and Sylvia Wynter, reads the ocean and its marine mammals for their history, their present, and their future. The legend of Drexciya and Afrofuturism come to mind. We discuss both in the course. Gumbs's volume is a meditation on race, on feminism, on multispecies relations, on entanglements, on environmental degradation, and on climate change and its impacts. We discuss these threads and strands. I bring in the relevant environmental science. Having read the texts earlier this semester, the students can more readily parse the deep history that structures *Undrowned*, even if it is not named. We discuss the text as one that acknowledges both the history that haunts and the current impacts of climate change. We also discuss how the book provides, without sidestepping that history, models, and celebrates Black life and Black livingness—livable lives. Given the numerous environmental and health risks we are currently navigating in our modern world, models for the livable that navigate the turbulence will be all the more relevant.

Notes

1. Epeli Hau'ofa's "Our Sea of Islands" argues for reading the relations among islands in the Pacific.

2. On land versus Land, see Liboiron 6–7n19.

3. This point is elaborated over the course of her oeuvre.

4. Philip references her earlier *Looking for Livingstone: An Odyssey of Silence.*

5. McKittrick discusses Black livingness in many forums, oral and written (see "Mathematics Black Life," "Diachronic Loops," *Dear Science*, "On Algorithms and Curiosities," and "On Black Methodologies").

Works Cited

Brand, Dionne. *At the Full and Change of the Moon.* Grove Press, 2000.

———. *A Map to the Door of No Return: Notes to Belonging.* Doubleday Canada, 2001.

———. "Nothing of Egypt." *Bread Out of Stone: Recollections, Sex, Recognitions, Race, Dreaming, Politics,* by Brand, Coach House Press, 1994, pp. 99–114.

Chakrabarty, Dipesh. "The Climate of History: Four Theses." *Critical Inquiry*, vol. 35, no. 2, winter 2009, pp. 197–222.

Davis, Heather, and Zoe Todd. "On the Importance of a Date; or, Decolonizing the Anthropocene." *ACME: An International Journal for Critical Geographies*, vol. 16, no. 4, 2017, pp. 761–80.

Glissant, Édouard. *Poetics of Relation.* Translated by Betsy Wing, U of Michigan P, 1997.

Gumbs, Alexis Pauline. *Dub: Finding Ceremony.* Duke UP, 2020.

———. *M Archive: After the End of the World.* Duke UP, 2018.

———. *Spill: Scenes of Black Feminist Fugitivity.* Duke UP, 2017.

———. *Undrowned: Black Feminist Lessons from Marine Mammals.* AK Press, 2020.

Haraway, Donna. *Staying with the Trouble: Making Kin in the Chthulucene.* Duke UP, 2016.

Hau'ofa, Epeli. "Our Sea of Islands." *The Contemporary Pacific*, vol. 6, no. 1, 1994, pp. 148–61.

Immerwahr, Daniel. *How to Hide an Empire: A History of the Greater United States.* Macmillan Publishers, 2019.

Lewis, Simon L., and Mark A. Maslin. "Defining the Anthropocene." *Nature*, vol. 519, 2015, pp. 171–80.

Liboiron, Max. *Pollution Is Colonialism.* Duke UP, 2021.

Lorde, Audre. "Grenada Revisited: An Interim Report." *The Struggle for Grenada*, special issue of *The Black Scholar*, vol. 15, no. 1, Jan.-Feb. 1984, pp. 21–29.

McKittrick, Katherine. *Dear Science.* Duke UP, 2021.

———. "Diachronic Loops / Deadweight Tonnage / Bad Made Measure." *Cultural Geographies*, vol. 23, no. 1, 2016, pp. 3–18.

———. "Mathematics Black Life." *The Black Scholar*, vol. 44, no. 2, summer 2014, pp. 16–28.

———. "On Algorithms and Curiosities." Keynote for Feminist Theory Workshop, Duke U, 8 May 2017.

McKittrick, Katherine, et al. "On Black Methodologies (and Other Things): A Conversation with Katherine McKittrick, Krista Franklin, and Alexander G. Weheliye." Intellectual Publics, City U of New York Graduate Center, 19 Apr. 2021.

Moore, Jason. Introduction. *Anthropocene or Capitalocene? Nature, History and the Crisis of Capitalism*, edited by Moore, PM Press, 2016, pp. 1–11.

Philip, M. NourbeSe. *Zong!* Wesleyan UP, 2011.

Roberts, Brian Russell, and Michelle Ann Stephens. "Archipelagic American Studies: Decontinentalizing the Study of American Studies." *Archipelagic American Studies*, edited by Roberts and Stephens, Duke UP, 2017, pp. 1–56.

Sharpe, Christina. "In the Wake: A Salon in Honor of Christina Sharpe." Barnard College, 2 Feb. 2017.

———. *In the Wake: On Blackness and Being.* Duke UP, 2016.

Spillers, Hortense. "Mama's Baby, Papa's Maybe: An American Grammar Book." *Diacritics*, vol. 17, no. 2, summer 1987, pp. 65–81.

Wynter, Sylvia. "1492: A New World View." *Race, Discourse, and the Origin of the Americas: A New World View*, edited by Vera Lawrence Hyatt and Rex Nettleford, Smithsonian Institution Press, 1996, pp. 5–57.

———. "On How We Mistook the Map for the Territory, and Reimprisoned Ourselves in Our Unbearable Wrongness of Being, of Desêtre: Black Studies toward the Human Project." *Not Only the Master's Tools: African American Studies in Theory and Practice*, edited by Lewis Gorden, Paradigm, 2006, pp. 107–69.

Yusoff, Kathryn. *A Billion Black Anthropocenes or None*. U of Minnesota P, 2018.

Clare Echterling

Decolonizing Climate Knowledge: Kathy Jetñil-Kijiner's Poetry

As a PhD candidate and then a lecturer at the University of Kansas (KU) between 2016 and 2020, I regularly taught an accelerated, online version of Global Environmental Literature, an upper-level undergraduate course that introduces students to global environmental issues, activism, and storytelling through ecocritical literary studies. In this eight-week, half-semester class, my students and I studied literature that reveals the connections between past and present imperialism, social injustice, and environmental problems in colonial and postcolonial contexts.

For the 2019–20 academic year, I redesigned Global Environmental Literature to incorporate a unit on the literature of climate change, which I had not previously addressed in the class. Because of the course's breakneck pace, I wanted to anchor the unit with short, approachable texts that would resonate with students while facilitating discussions about climate justice, the Anthropocene, and anti-colonial climate literature. Consequently, I chose to focus the course's second unit on the Marshall Islander poet, performer, and activist-educator Kathy Jetñil-Kijiner, whose poetry and public speaking performances make visible the colonial roots of the climate crisis and the necessity of climate justice. In this essay, I discuss my approach to teaching Jetñil-Kijiner's poetry—specifically "Dear Matafele Peinam," "Two Degrees," and "Rise," a poem coauthored by the Inuk writer Aka Niviâna—in an online format to students from a variety of academic disciplines and reflect on their responses to the unit.

Course Goals and Structure

Over the years that I taught it, Global Environmental Literature attracted students from across a wide range of programs at KU's flagship and satellite campuses. Cross-listed in English, environmental studies, and global and international studies, the course unsurprisingly drew many students from those departments. However, it also counts toward KU's Global Awareness Program Certificate, which students from any program can pursue. It furthermore satisfies the global awareness and cultural understanding component of the required undergraduate core curriculum. Consequently, students from across the university came into the course with very different prior knowledge, perspectives, and needs. This diversity led to dynamic class discussions and ultimately influenced my goals for the course and the way I taught it.

The students did share an interest in the environment. Nearly everyone who enrolled in the course already cared about environmental issues, but most started the class with parochial, mainstream understandings of environmentalism and little to no knowledge of anti-colonial environmental literature, environmental justice, or what both Joan Martinez-Alier and Ramachandra Guha call "the environmentalism of the poor." My goal was thus to begin decolonizing students' knowledge of environmental issues, activism, and literature. In other words, I wanted to open students' eyes to the environmental injustices created by ongoing imperial oppression, introduce them to anti-colonial environmental justice movements in the Global South, and help them think critically about the role of literature in the struggle for just and sustainable futures.

I divided the asynchronous course into seven units, which included five one-week units, one two-week unit, and one week to complete a final paper. In the first six units, I assigned one or more primary texts alongside secondary sources, videos, and short prerecorded lectures that introduced key concepts and provided crucial historical, political, and ecological context for the primary text or texts. Because many of the students were new to literary criticism or to environmental studies, at the beginning of each unit I sent out a reading guide that highlighted central themes and questions to help them focus their engagement with the material and prepare for upcoming assignments.

Each unit included a reading response and a class discussion. The former called for mini-essays in response to questions about the unit's primary and secondary texts. To ease student stress, I made it possible to revisit and revise the responses over time as long as final versions were submitted before the end of the unit. Discussions were similarly flexible. After a few semesters struggling to generate good conversation on *Blackboard* discussion boards, I moved discussions to *Kami*, a free web tool that allows users to annotate files collaboratively. In each unit, I chose one or two of the primary or secondary readings as an anchor for our conversation. I would start the conversation by posting a few questions. Students then had two weeks to complete a required number of posts, which gave them ample time to complete the reading and participate in the discussion. It likewise gave me plenty of time to read their posts, offer feedback, and steer the conversation. I found this discussion format to be very successful. By writing directly on a text, students were able to practice close reading and help one another analyze difficult passages. This format also fostered higher student engagement and more exciting conversations than traditional discussion boards. Ultimately, both the *Kami* discussions and the reading responses helped students engage with the material and develop their literary analysis skills before completing the course's major assignments, a midterm essay and a final paper. Furthermore, they allowed me to support students through written feedback as they learned about and grappled with global environmental injustice and the literature that bears witness to it.

Unit 1: Foundations

Global Environmental Literature's first unit focused in part on Ken Saro-Wiwa's detention diary, *A Month and a Day*. More importantly for the topic of this essay, it also laid the foundation for the course by introducing key concepts and issues that we revisited throughout the term, including environmental justice, the environmentalism of the poor, "slow violence" (Nixon, *Slow Violence*), and the role of environmental literature. It thus prepared students to learn about climate justice and Jetñil-Kijiner's writing in the second unit.

On the first day of class, I sent out the unit's reading guide. Since so many of my students were relatively unfamiliar with literary studies, I included in the guide an introduction to ecocriticism. I also provided other resources on critical reading, writing about literature, research, and MLA style that students could reference throughout the course. To supplement a lecture on Saro-Wiwa and excerpts from *A Month and a Day*, I assigned secondary readings that together created a critical framework for students to analyze the literature we would read. These included Martinez-Alier and Giovanna Di Chiro's respective essays in *Keywords for Environmental Studies*, part of chapter 6 from Guha's *Environmentalism: A Global History* (98–108), and Rob Nixon's introduction to *Slow Violence and the Environmentalism of the Poor* (1–44).

The unit's *Kami* discussion focused on Nixon, although students also drew connections to the other texts. In the assigned essay, Nixon develops his theory of slow violence, defining it as "violence that occurs gradually and out of sight" (*Slow Violence* 2). Students readily grasped the concept, although many acknowledged that they had never thought of violence in this way or, for that matter, of environmental degradation as violence at all.

Nixon also helped orient students to ecocritical ways of reading and the potential of environmental justice literature. In the reading guide, I asked students to pay close attention to Nixon's points about the role of the writer-activist and the ability of literature to make visible the slow violence of environmental injustice. On *Kami*, students consequently zeroed in on these aspects of the author's argument, discussing how literature and other forms of storytelling can represent slow violence. Ultimately, they would agree with Nixon's assertion that the environmental writer-activist's task is to show people slow violence and move them to action. Since the first *Kami* discussion spilled over into unit 2, students were still thinking about this idea as we turned our attention to Jetñil-Kijiner and climate injustice in the Anthropocene.

Unit 2: Climate Justice, the Anthropocene, and Jetñil-Kijiner's Climate Poetry

Indigenous, decolonial, and postcolonial scholars have illustrated in recent years that climate change and the Anthropocene stem from the violent processes of colonialism and capitalism (Davis and Todd; Whyte). Moreover, as the Potawatomi scholar Kyle Whyte argues, climate change and the other crises of the Anthropocene are a continuation and "intensification" of the ecological changes caused by colonialism (153, 154–57). Furthermore—as climate justice activists and scholars have pointed out—poor and marginalized communities are the least responsible for the climate crisis but are nevertheless the most vulnerable to it. For the unit on climate literature, therefore, I wanted to show students that climate change and the Anthropocene are rooted in empire while introducing them to climate justice. My other objective, building from the first unit's focus on the role of environmental justice literature, was to help them recognize the importance of narratives that make legible the inequalities of the climate crisis.

To accomplish this, I provided a ten-minute introductory lecture on climate literature, the Anthropocene, and Jetñil-Kijiner's work, which I asked students to watch before starting the unit's texts. Students then read a short definition of climate justice from *The Routledge Handbook of Climate Justice* (Jafry) and Nixon's essay on vulnerability, economic inequality, and Anthropocene storytelling. Nixon calls for stories that expose the injustices of the Anthropocene, or, as he puts it, "the way the geomorphic powers of human beings have involved unequal exposure to risk and unequal access to resources" ("Great Acceleration").

Finally, students watched the keynote speech at the 2018 Innovative City Forum, in which Jetñil-Kijiner details the history of American imperialism in the Marshall Islands and the impact of sea-level rise on the nation ("Kathy Jetñil-Kijiner"). Between 1946 and 1958, the United States conducted sixty-seven nuclear tests in the Marshall Islands, destroying Marshallese homelands, forcing migration, and subjecting citizens to the devastating health consequences of irradiation. Although largely forgotten in the United States, the aftereffects of the testing program continue to impact Marshallese life today. Now, rising oceans threaten to make the islands uninhabitable—another type of colonial theft.

Along with the previous unit's reading, this set of supporting materials prepared students to analyze Jetñil-Kijiner and Niviâna's poems. To keep the workload manageable, I assigned only three poems: "Dear Matafele Peinam" and "Two Degrees," which both appear in Jetñil-Kijiner's collection *Iep Jāltok: Poems from a Marshallese Daughter* (70–73, 76–79), and "Rise," which Jetñil-Kijiner and Niviâna wrote for a performance film produced by 350.org and directed by Dan Lin. Students watched Lin's film, *Rise: From One Island to Another*, which features Jetñil-Kijiner and Niviâna reading the poem in their homelands, Aelon

Kein Ad (the Marshall Islands) and Kalaalit Nunaat (Greenland). They also viewed Jetñil-Kijiner's performances of "Dear Matafele Peinam" at the opening ceremony of the United Nations secretary-general's 2014 climate summit and of "Two Degrees" during the 2016 UN climate summit ("Statement"; "Marshall Islands").

Students readily recognized the poems as galvanizing calls for climate justice—for the Marshall Islands and for other marginalized communities at the front lines of the climate crisis. In their reading responses and on *Kami*, students analyzed the ways the poems shed light on the slow violence of climate change and the ways imperial structures make oppressed populations worldwide particularly vulnerable to its effects. Relatedly, they argued that the poems fulfill Nixon's requirements for creative writing that exposes and resists the injustices of the Anthropocene. As my student Alexis Rawlings put it in her reading response, the poems are "the type of literature that Nixon wants to see in regard to the Anthropocene conversation and . . . can help to spread awareness of these issues to people who may otherwise not realize that people are already being impacted by climate change." Furthermore, she asserts, "This [type] of literature is necessary . . . to provide a sort of shock factor to encourage action."

In general, students were particularly struck by the emotional power of the poetry performances. In their reading responses, for example, many discussed how the footage of Jetñil-Kijiner and Niviâna reading "Rise" in their ancestral landscapes amplified the poem's indictment of the "colonizing monsters" who have stolen and contaminated indigenous land and are driving climate change (line 97). Similarly, many wrote about the significance of Jetñil-Kijiner's testimony and performance of "Dear Matafele Peinam" at the 2014 UN climate summit. Several pointed out that the performance is an example of speaking truth to power and a demand for wealthy nations to recognize and take responsibility for the climate crisis that imperils the Republic of the Marshall Islands and other frontline communities.

Alongside the poems, Jetñil-Kijiner's speech at the Innovative City Forum helped students see that climate change is part of a long legacy of imperial violence in the Marshall Islands. Most of the American students had never heard of the United States' nuclear testing program there and understandably struggled with that knowledge, variously expressing anger, horror, and embarrassment at the nation's cruelty. A few astutely pointed out that the United States had transformed the Marshall Islands into a "sacrifice zone" and that now—because of wealthy nations' failure to

act on climate—the islands are being sacrificed again. This "long history
of misuse," as one student, Jessica Blom, put it, proved to many of them
that imperial powers saw and continue to see the Marshallese and their
homelands as disposable.

Many students also focused on how "Dear Matafele Peinam" and
"Two Degrees" especially illustrate the necessity of safeguarding the climate for future generations and the continuation of Marshallese culture.
In "Dear Matafele Peinam," for example, Jetñil-Kijiner addresses her in-
fant daughter, promising her that

> no one's losing
> their homeland
> no one's gonna become
> a climate change refugee
>
> or should i say
> no one else (lines 37–42)

On *Kami,* students pointed out how this poem argues that Jetñil-Kijiner's
daughter and the Marshallese "deserve" not "just / [to] survive" but "to
thrive" (88, 90–91, 93).

Relatedly, most found Jetñil-Kijiner's invocation of family particularly moving. In "Two Degrees," the poet compares global warming to a
toddler's fever, meditating on the huge impact a slight fever can have on a
child. Likewise, the poem explains, a global temperature increase of just
two degrees Celsius will drown the Marshall Islands. Students especially
liked the poem's reminder

> that beyond
> the discussions
> numbers
> and statistics
> there are faces (100–04)

As many observed, these lines are a reminder that climate change is not
an abstract problem but a present danger to people living now.

In fact, nearly every student remarked on how the poems humanize
climate change and make it personal, thus helping them understand the
issue in a way that statistics and data never had. Several remarked that the
poems were the most impactful thing they had ever read about climate
change. Their emotional responses brought them back to discussions about
the vital role of climate justice literature. While some students struggled

with feelings of helplessness and guilt, many remarked that the poems motivated them to take action and become allies of the communities who are most at risk from climate change. Consequently, several students observed, it is important for people in positions of privilege to encounter and sit with stories of climate injustice.

Given my previous experiences teaching Global Environmental Literature, when I redesigned the class to include a unit on climate literature, I suspected that most students would begin the term knowing little to nothing about climate injustice and the connections between imperialism and climate change. Indeed, in the four sections I taught during the 2019–20 academic year, student writing confirmed that most—including some environmental studies majors—began the course thinking that climate change affects everyone more or less equally and that everyone is more or less equally responsible for it. However, their writing also demonstrated that the unit drastically changed their thinking about climate change, showing them that addressing the climate crisis necessarily involves dismantling imperial power structures. As multiple students pointed out, the unit made them understand that climate change is a human rights issue, not just a problem that affects polar bears or coral reefs. As I had hoped, most students came away from the unit demonstrating a keen understanding that climate change does not in fact affect us all equally and that the world's most marginalized communities are the most vulnerable to it.

Works Cited

Adamson, Joni, et al., editors. *Keywords for Environmental Studies.* New York UP, 2016.

Davis, Heather, and Zoe Todd. "On the Importance of a Date, or Decolonizing the Anthropocene." *ACME: An International Journal for Critical Geographies*, vol. 16, no. 4, 2016, pp. 207–21.

Di Chiro, Giovanna. "Environmental Justice." Adamson et al., pp. 100–05.

Guha, Ramachandra. *Environmentalism: A Global History.* Longman, 2000.

Jafry, Tahseen, et al. "Introduction: Justice in the Era of Climate Change." *The Routledge Handbook of Climate Justice*, edited by Tahseen, e-book ed., Routledge, 2018.

Jetñil-Kijiner, Kathy. *Iep Jāltok: Poems from a Marshallese Daughter.* U of Arizona P, 2017.

Jetñil-Kijiner, Kathy, and Aka Niviâna. "Marshall Islands Poet to U.N. Climate Summit: At Two Degrees, My Islands Will Already Be under Water." *YouTube*, uploaded by Democracy Now!, 18 Nov. 2016, www.youtube.com/watch?v =SdqyInZQb28.

———. "Rise." *350.org*, 350.org/rise-from-one-island-to-another/.

"Kathy Jetñil-Kijiner 'Island Cities under Water: The Case for Marshall Islands'—Keynote 3." *YouTube*, uploaded by Innovative City Forum, 13 Nov. 2018, www.youtube.com/watch?v=cGsN7tQ_R24.

Martinez-Alier, Joan. "Environmentalism(s)." Adamson et al., pp. 97–100.

Nixon, Rob. "The Great Acceleration and the Great Divergence: Vulnerability in the Anthropocene." *Profession*, Mar. 2014, profession.mla.org/the-great-acceleration-and-the-great-divergence-vulnerability-in-the-anthropocene/.

———. *Slow Violence and the Environmentalism of the Poor*. Oxford UP, 2011.

Rise: From One Island to Another. Directed by Dan Lin. *Vimeo*, uploaded by Mainspring Media, 2018, vimeo.com/289482525.

Saro-Wiwa, Ken. *A Month and a Day: A Detention Diary*. Penguin Books, 1995.

"Statement and Poem by Kathy Jetnil-Kijiner, Climate Summit 2014—Opening Ceremony." *YouTube*, uploaded by United Nations, 23 Sept. 2014, www.youtube.com/watch?v=mc_IgE7TBSY.

Whyte, Kyle. "Indigenous Climate Change Studies: Indigenizing Futures, Decolonizing the Anthropocene." *English Language Notes*, vol. 55, nos. 1–2, 2017, pp. 153–62.

Mary Ann Gosser-Esquilín

Sounding the Alarm of Climate Change in Caribbean Literature: Mayra Montero's *In the Palm of Darkness*

Teaching Caribbean literary texts related to climate change is an ethical imperative because for islanders the effects of deforestation, disappearing species, and rising sea levels have become matters of life and death. Furthermore, ecological degradation has punctuated the losses of traditions and culture intimately related to the natural world. From its insertion into the world's geopolitical vortex in 1492, the Caribbean ecosystem was set on a course that has brought on the environmental devastation now assailing the region. As the colonial enterprise of gold mining became unprofitable on the islands, growing sugar provided great wealth but exacted enormous costs. The hard labor and environmental destruction involved in clearing forests; planting, tending, and harvesting sugarcane; and then refining raw sugars devastated native peoples and their habitats. Having thus reduced the population upon whom it relied for indentured labor, "[t]he system of plantation agriculture at the centre of the sugar revolution came quickly to an almost exclusive dependence on a single form of labour and a particular group of people—enslaved Africans and their Caribbean-born children" (Higman 122). The massive importation of people, their forced labor, and the cultivated crops have permanently altered insular landscapes. Caribbean authors acknowledge that, given these dramatic transformations of the ecosystem, the world's ecological destinies have become inextricably bound together as climatic changes accelerate.[1]

Teaching Mayra Montero

In 1995, the Cuban Puerto Rican novelist and journalist Mayra Montero published the region's first econovel, *Tú, la oscuridad* (*In the Palm of Darkness*).[2] The plot centers on the quest of a herpetologist from the United States, Victor Grigg, and his Haitian guide, Thierry Adrien, Jr., for what will turn out to be the last specimen of the *Eleutherodactylus sanguineus* frog (locally known as the *grenouille du sang*, or "blood frog"). Montero decided to write this novel, her third one set in Haiti, after reading about the sinking of an overcrowded ferry, the *Neptune*, leaving the port city of Jérémie on its way to Port-au-Prince, Haiti, in February 1993. Her protagonists board that ferry in order to flee, as quickly as possible, the death threats received on a deforested mountain. Throughout the novel, the desertification and drought that afflict the Haitian landscape translate into the loss of several species and serve as backdrops to political and drug-trafficking assassinations in the poorest nation of the Western Hemisphere.[3]

The novel serves as the central and longest literary text when I teach an interdisciplinary first-year honors course (Introduction to Caribbean and Latin American Studies) that fulfills one of the two Foundations in Global Citizenship courses required by the general education curriculum of our university's Intellectual Foundations Program. This requirement helps students develop an "awareness of global connectedness and interdependence, understanding how their actions can affect other peoples and places" ("Intellectual Foundations").[4] Throughout the course, we use the most recent edition of *Modern Latin America*, by Peter H. Smith and James N. Green, which contains only one Caribbean case study: Cuba. We therefore present other Caribbean nations through literary texts that provide an artistic means to illustrate the history of the area. Because we do not have the luxury of grounding *In the Palm of Darkness* within a Caribbean literary tradition, I encourage students to watch the Cuban film *Cumbite* on *YouTube* ("Cumbite"). The film, based on the Haitian author Jacques Roumain's 1944 novel *Gouverneurs de la rosée* (*Masters of the Dew*), deals with the desertification of Haiti and the aftermath of the United States occupation (1919–34) and sets the stage for the more noticeable climate change consequences depicted in Montero's work decades later.

I also assign the documentary *Extinction in Progress*, which offers current, impactful visuals and sobering figures depicting climate change in Haiti; for example, the forests that covered thirty percent of the nation's land in 1940 had shrunk to ten percent of the land by 1970 and had fallen below two percent by 2013 ("Extinction" 3:38–52). The scientists in the film travel to remote locations high up in the mountains to investigate the state of the biota's diversity. To reach those distant peaks, they first travel through towns and local markets and focus on charcoal, the ubiquitous domestic energy fuel of many Haitians. The documentary shows great quantities of charcoal juxtaposed with the mountain forests that were stripped of trees to make it while smoke lingers everywhere. The stark imagery establishes a correlation between these realities to signal how charcoal burning is partly responsible for greenhouse gas emissions, which accelerate climate change.

Historical Overviews

To teach about Haiti, even in a diverse, four-year public university in southeast Florida, one must provide a historical overview of Saint-Domingue, France's richest colony at the onset of the Haitian Revolution in 1791. Many students (including those from the Caribbean) do not know much about Haiti's past, its struggle for the abolition of slavery, its road to declaring independence in 1804 and to gaining recognition as a new republic (the second in the Americas after the United States and the only nation established through a successful revolt by enslaved people), and the impact racial tensions and political events have had on the environment. Through figures and tables, students learn that by 1780, "there were 28,000 whites, 22,000 *affrinchis* and free people of color, and 460,000 black slaves" in Saint-Domingue (Figueredo and Argote-Freyre 85).[5] Jon Rogoziński's *A Brief History of the Caribbean* also provides sobering numbers for Saint-Domingue. Between 1761 and 1810, the number of enslaved people stands at 481,000. Rogoziński does not record any more because both the 1801 and 1805 constitutions explicitly abolish slavery forever in Haiti (124). Eventually, Haiti became a nation of small farmers: "Some owned their own plots and others became sharecroppers . . . There [in the mountains] they cleared the forests and opened the land to massive soil erosion" (220).[6] Students become aware that deforestation punctuates the effects of slow climate change, which since the declaration of independence have been exacerbated by internecine divisions, the weighty indemnity debt (150 million francs) owed to France, and the United States' fears of European interventions in the Americas as its economic interests in Haiti (e.g., logging) grew during its occupation of the country.[7]

I provide this historical, political, economic, and ecological overview of Haiti's past during sessions prior to the discussion of the novel. I encourage students to consider that enslaved people and their descendants, like Haiti's animals and plants, were victims of anthropogenic decisions (e.g., planting sugarcane) made with no concern for them or their cultures of preservation by those who viewed them as nonhuman Others—an exploitative mindset that endures to this day. Next, we review Haiti's situation during the time frame of the novel. Victor arrives in Haiti in October 1992. He disappears, together with Thierry and the last blood frog, when the *Neptune* sinks on 16 February 1993. Significantly, that span of time also represents the duration of the first voyage of Columbus to the Caribbean, five hundred years earlier, up to the date of his first letter (15 February 1493)—to the Spanish Crown, in which he marvels at the lush vegetation and the docility of the people he encounters during his first voyage. Victor, as a modern version of the Genoese explorer, cannot extol the richness of the landscape or take live specimens, because exploitations of the ecosphere have ensured that these no longer exist. In fact, he finally acknowledges, "Haiti was disappearing" (Montero, *In the Palm* 171).

By February 1993, Jean-Bertrand Aristide, Haiti's first democratically elected president, ousted by a coup d'état led by the military in September 1991, sought to return to power. The violence to which the novel alludes (political persecutions, murders, and the torture of Aristide supporters, combined with the brutality of the drug lords abetted by the military) relates to the aftermaths of the coup. How, we ask in class, can Haitians and a Haitian government in disarray concern themselves with climate change when their everyday lives and subsistence remain precarious?

We also study a map (the English translation of the novel comes with a topographical map noting cities, mountains, and locations visited by the characters). The helpfulness of a map cannot be overstated: one appreciates the difficulty of the terrain, the proximity to Cuba, and the shared border with the Dominican Republic.[8]

Structure and Discussion of the Novel

Before students read *In the Palm of Darkness* (which I assign in three parts), we spend time discussing the structure of the novel, which consists of twenty titled chapters. The odd-numbered ones are told from Victor's perspective, in a stream-of-consciousness style. The even-numbered chapters present Thierry's history and offer the insights of a Vodou practitioner and member of a male secret society into the significance of the biota to the religiosity of Haitians. However, in between those chapters, nine untitled, third-person-omniscient scientific vignettes or notes record the disappearance of frogs throughout the world, thus pointing to the effects of climate change and the urgency of recording the demise of vanishing species.[9] These notes list the names of reports and scientists together with the Linnaean taxonomic names of the lost amphibians and complement the actions of the fictional characters by underscoring that the Haitian frog's case represents one of many worldwide. Nestled between the protagonists' chapters, they remind us that the disappearance of frogs, even in wealthy nations promoting environmental policies, sounds the alarm of the effects of slow climate change.

Once we start the novel, students encounter the voice of another Haitian, Dr. Emile Boukaka, "a surgeon and an amateur herpetologist" (91) and a member of the brotherhood, who attributes the demise of the frog and its habitat not only to drought and deforestation but also to sociocultural abuses of the environment: the hills are being used to dispose of bodies (either for political or drug-trafficking gain). The doctor, who has published articles on the decline of amphibians, exclaims, "'The great flight has begun. . . . You people invent excuses: acid rain, herbicides, deforestation. But the frogs are disappearing from places where none of that has happened'" (96). Our discussions of this quotation focus on this view from a doctor in Haiti, in which the majority of the population descends from enslaved Black Africans, as opposed to external climate change deliberations related to Haiti, which should consider racial politics and economics.

Course Assignments

Students, divided into groups of two or three and assigned a Caribbean or Latin American nation to research, create two wikis for their respective countries. The first wiki has to contain basic factual information as well as an item the contributing group found surprising. The second wiki presents a recent news item and a reflection as it relates to material learned in class about the assigned country. Students review and critique all wikis produced in the class. Haiti is not among the assigned countries, and prior to reading the Montero novel, students are asked, "What do you know about Haiti?" Many admit to knowing little or nothing; others mention food, Vodou, poverty, and natural disasters, such as hurricanes and the 2010 earthquake.

A project I had planned called Mapping Disappearances evolved, amid the difficulties of transitioning to remote teaching and learning, into a short research and reflection paper (250 words).[10] Students look up a recent article (and provide documentation) addressing the disappearance of a species or any other ecological concern of their assigned country and offer a reflection in the light of what they have learned from the novel and the documentary. The reflection can be of a personal nature. A recurring concern has turned out to be that COVID-19 will displace climate change as a subject of discussions and funding.

The last project related to the novel, Voices of Others, works well as a discussion board. Each student is responsible for creating an original post (200 words), as well as two replies (100 words each) to classmates' posts, as a character from *In the Palm of Darkness*. The characters are chosen to represent different perspectives on the urgency of the disappearance of the frog, and each student, through their selected character, offers insights that stem from their character's circumstances. They are asked to write as Ganesha, a poor Indo-Guadeloupean living in Haiti; the female cactus, which is becoming extinct; Julien, a former *macoute*[11] turned hired killer; Yoyotte Placide, a cook and one of Thierry's father's lovers; or Barbara, a geologist from the United States and lesbian lover of Victor's wife, in the following terms: "Provide their background (class, race, gender); then imagine how they would react to the disappearance of the *grenouille du sang*. What would be their concerns? Would they have other concerns? To whom would they address them? How and why?" At first, this may seem like a simplistic exercise, but for those students who engaged with their character, especially the nonhuman Other of the female cactus, the effect on those who reacted to the posts became noticeable. One of the students asked during class, "How does one assume the persona of a cactus?" I explained that they had to use their imagination and think creatively. They decided that the idea—giving voice to a cactus, a life form often not highly esteemed for the role it plays in the biosphere—was worth the challenge and gave them a venue to present inventively and apply what they had learned about climate change in the novel to very dramatic effects. In the student's post, the cactus addresses the frog as sharing a similar fate and concludes, "I hope both of our species are found by those who seek us, so that our respective kind is saved from vanishing."

Two students volunteered that they would have appreciated two additional tools for reading the novel: a glossary and translation of Haitian words or concepts and a character chart to help with the intricate relationships over various decades among the many disappeared characters. These valid suggestions will be implemented.

Urgency of the Work

In the Palm of Darkness offers a unique Caribbean perspective on Haiti's environmental struggles and the many forces at play in the making of its ecological crisis. The novel provides a jarring warning of what lies ahead by sounding the alarm on how slow climate changes lead to the disappearance of nonhuman Others and their cultural significance. The work invites us to consider the effects of these changes on a nation at the epicenter of Caribbean history.

Notes

1. For further information about clearing the land for sugarcane and about the number of enslaved people brought to the Caribbean to work, see Gibson 95–125. For information on slave ships, maps, and the transportation of enslaved people, visit *Slave Voyages: The Trans-Atlantic Slave Trade Database* (slavevoyages.org).

2. The Caribbean ecoliterary critic Lizabeth Paravisini-Gebert asserts that "*In the Palm of Darkness* is an avowedly environmentalist novel—the region's first" (192).

3. At the same time, Sarah, a botanist from the United States, searches for the female specimen of a cactus (*Pereskia quisqueyana*) with the same intensity as Victor. The novel hints that since she does not heed the warnings to leave, she will become another victim at the hands of those who control the mountain.

4. I have also taught the novel in a comparative Caribbean literature graduate seminar.

5. The term *affranchis* (spelled "affrinchis" in the quotation) describes "the freed offspring of a white and black or free-born people of color. This latter term also referred to African slaves who earned their freedom as well as mulattoes" (Figueredo and Argote-Freyre 85).

6. Rogoziński's figures reveal that the people trade and furthering of the Caribbean's ecological devastation found a new home in Cuba: between 1761 and 1810, 139,200 enslaved people were imported; between 1811 and 1870, the number explodes to 550,000 (124). Funes Monzote provides a detailed study of the relationship between the abolition of slavery and decline in sugar production in Haiti and the expansion of slavery and sugar production in Cuba after Haiti's declaration of independence.

7. France agreed to recognize Haiti as a republic in exchange for this payment in 1825. The United States, fearing that acknowledging Haiti's independence would encourage enslaved people in America to revolt, did not do so until 1862.

8. The relationship between the two nations remains tense. Haitians invaded the Dominican Republic in 1822 and occupied it until 1842, and the memory of the 1937 massacre of Haitian sugarcane cutters alongside the border persists.

9. The vignettes encompass research done from the 1930s to as recently as 1993 in Colorado, Australia, Switzerland, Costa Rica, Hawai'i, Puerto Rico, Colombia, California, and Honduras.

10. Students would have used *Google Maps* to enter the locations of the frog disappearances as noted in Montero's novel and of the disappearance of a species within their assigned wiki countries. The students would have reflected on how the disappearance they had researched affected their assigned country, its people, its politics, and its economics, and how this information influenced their views of that country or its environmental policies.

11. The Tonton Macoute, named for a folkloric figure of an "uncle" (*tonton*) with a burlap bag (*macoute*) for kidnapping ill-behaved children, became the most feared of François "Papa Doc" Duvalier's paramilitary forces.

Works Cited

Cumbite. Directed by Tomás Gutiérrez Alea, Instituto Cubano del Arte e Industria Cinematográficos, 1964.

"Cumbite (1964) aka Coumbite [Subs: English, Español]." *YouTube*, uploaded by j, 14 June 2015, youtube.com/watch?v=uWVeMa9LZsU.

"Extinction in Progress (Documentary)." Directed by Jürgen Hoppe, 2013. *YouTube*, uploaded by Blair Hedges, 18 Nov. 2019, www.youtube.com/ watch?v=Lo5Ls89pJ_4.

Figueredo, D. H., and Frank Argote-Freyre. *A Brief History of the Caribbean.* Facts on File, 2008.

Funes Monzote, Reinaldo. *From Rainforests to Cane Field in Cuba: An Environmental History.* 2004. Translated by Alex Martin, U of North Carolina P, 2008.

Gibson, Carrie. *Empire's Crossroads: A History of the Caribbean from Columbus to the Present Day.* Grove Press, 2014.

Higman, B. W. *A Concise History of the Caribbean.* Cambridge UP, 2011.

"Intellectual Foundations Program (IFP) for Students." *Florida Atlantic University*, 2023, fau.edu/ugstudies/ifp-curriculum-sheets/.

Montero, Mayra. *In the Palm of Darkness.* Translated by Edith Grossman, HarperCollins Publishers, 1997.

———. *Tú, la oscuridad.* Tusquets Editores, 1995.

Paravisini-Gebert, Lizabeth. "'He of the Trees': Nature, the Environment, and Creole Religiosities in Caribbean Literature." *Caribbean Literatures and the Environment: Between Nature and Culture*, edited by Elizabeth DeLoughrey et al., U of Virginia P, 2005, pp. 182–96.

Rogoziński, Jan. *A Brief History of the Caribbean: From the Arawak and the Carib to the Present.* Facts on File, 1992.

Smith, Peter H., and James N. Green. *Modern Latin America.* 9th ed., Oxford UP, 2018.

Nassim W. Balestrini

The Polymedial Aesthetics
of Climate Change Drama

Literary studies classes on ecocriticism and environmental justice benefit greatly from incorporating drama. Stage representations of a long-term and large-scale phenomenon like climate change encourage students to think outside the communicative box while considering questions like the following: How does theater convey thoughts about the effects of climate change on the planet, specific environments, and interactions among living beings? How can the limited time frame of a performance accommodate a geological age like the controversially named Anthropocene? As I want my students to think about how playwrights accommodate the immense timescale of climate change within performances of specific lengths, my teaching strategy combines a full-length play with several five-minute microdramas. I used this approach in two seminars at the University of Graz, Austria, for master's students in English and American studies and English as a second language. Seminars in literary and cultural studies at the university accommodate up to twenty-four students, comprise fourteen ninety-minute sessions, and are taught in English.

I chose to teach the Canadian-born, New York–based playwright Chantal Bilodeau's full-length play *Sila*, whose Inuktitut title references the Inuit conceptualization of an all-encompassing breath of life. The seminar participants analyzed Bilodeau's artistic employment of the dramatic characters' mobility on land and ice, in water, horizontally and vertically, between life and death, and between physical and metaphysical realms. For the sessions on microdramas, I selected works commissioned for Climate Change Theatre Action (CCTA). Having originated as a three-month project timed to coincide with the 2015 United Nations climate conference, CCTA has evolved into a biennial theater and activist initiative during which anyone can use its current collection of commissioned plays for climate-change-related events (Bilodeau, Introduction 17; Balestrini, "Anthropogenic Climate Change" and "Transnational and Postcolonial Perspectives"). Furthermore, I opted for a two-pronged pedagogical approach. First, students developed an online glossary of theories and methodologies for critical analysis of the printed dramatic texts. Subsequently, they engaged creatively with the plays from a theater director's vantage point.

Sila: An Inuit Perspective on the Breath of Life

Sila, the first play in a projected eight-part series entitled *The Arctic Cycle*, is set on Baffin Island, Canada, and depicts the impact of global warming and of the fossil fuel industry on the Canadian Arctic.[1] Interrelated plotlines emphasize experiences of loss and mourning that cut across ethnicities, nationalities, species, generations, and genders. Indigenous and white characters, polar bears, and mythological beings expose the entanglements and deadly impacts of profit-focused and racialized politics. Counterforces in the play are knowledge-based empathy and the experience of connectedness to the life force, or *sila*, which serves as a wellspring of physical survival and artistic articulations of climate change activism (Balestrini, "Climate Change Theater"; May 252–67).

So far, I have worked with two approaches to teaching *Sila*. The first option is to provide three areas of inquiry, each elaborated in secondary sources that center on relevant theories or present case studies. On ecocritical and postcolonial theory, I assign Cara Cilano and Elizabeth DeLoughrey, Greta Gaard, Linda Hutcheon, and Hubert Zapf; on the combination of human and animal characters on stage, I assign Robert Baker-White and Una Chaudhuri; and on anthropological findings on Inuit mobility, I assign Claudio Aporta. The other approach requires students to find secondary texts and explain why they consider their theoretical outlook useful. Either option allows students to explore Indigenous versus Western conceptualizations of mobility, relations between space and sociality, and the impact of mobility on geographically demarcated eco-systems, thus encouraging them to grapple with culturally embedded notions of selfhood, belonging, and positionality.

Some seminar participants identified linkages between postcolonial theory and environmental justice as helpful for studying *Sila*'s treatment of "displacement" and, "as a sub-topic," "language" as an indicator of "cultural displacement" or "powerlessness."[2] They also discussed place and belonging in *Sila* through conceptualizations of "home" and related these ideas to the play's representation of nonhuman characters without resorting to cute-ification or anthropomorphism. Those students who familiarized themselves with Inuit ideas and practices by reading Aporta's anthropological study on the social functions of trails in Arctic cultures felt suitably equipped to discuss Bilodeau's depiction of the Indigenous characters' predicament as a human rights issue. This experience also prepared seminar participants to contemplate *Sila*'s embedding of Indigenous knowledge within an otherwise Western theater format.

Many of these themes resurfaced when students were asked to identify and apply what they considered useful theoretical vantage points. Selecting Bilodeau's concept of interconnectedness (see Balestrini, "'Writing'" 36–39), one student invited her peers to unravel the role of interconnectedness in the character constellation, intersecting plotlines, and overall argument of *Sila*. In response, students emphasized the play's recurring parent-child relationships and experiences of mourning as connecting human characters of all ethnicities as well as human, animal, and mythical characters. They pointed out how Bilodeau's notion of "horizontal interconnectedness" (Balestrini, "'Writing'" 37) highlights analogies between physical mobility and communication.

Moreover, our discussion addressed intermedial strategies (Balme 205–08; Rajewsky) in *Sila* and the activist impetus toward breaking through the fourth wall. While theater per se usually comprises multiple semiotic channels (such as verbal, visual, and sonic signs), scrutinizing drama from an intermedial perspective inquires both into the perceived boundaries or distinctions between media and into their overlap, interaction, and mutual evocation. Students stressed that the use of words in theater extends beyond dramatic dialogue because *Sila* includes a formal speech, a letter, and the performance of spoken word poetry. When Veronica, the central Inuit character, loses her teenage son to suicide or accidental death by sniffing gasoline, she temporarily cannot access her life- and voice-bestowing *sila*. Through another character's unexpected and epiphanic empathy, she eventually regains the ability to speak and to perform poetry.[3] As they observed Veronica shifting attention away from herself to larger contexts through poems that confront racial prejudice and indicate possible ways into a better future, students recognized in the character's verbal prowess the communicative power of artistic climate-change activism. Other intermedial components are music (Balestrini, "Sounding") and projected images or words. Medium choices in representing the polar bears as life-size entities offer inroads into innovative climate change theater aesthetics with an activist trajectory—possibly by evoking a porous or torn fourth wall in order to appeal to audience members.

Climate Change Theatre Action

The second segment of studying climate change drama enables students to integrate their knowledge of ecocritical, postcolonial, and other theories into creatively imagining how they would direct a play. I provided them with questions as food for thought on issues like dramatic methods of communicating specific aspects of climate change; non-Western or Western traditions of characterization, character constellation, dialogue, sound, lighting, setting, and plot structure; polymedial and intermedial techniques; representations of science; realist and nonrealist features; social relations and conflicts; the impact of brevity; and conceptualizations of the Anthropocene, of "planetarity," and of human and nonhuman nature.[4] In response, seminar participants developed either a director's handbook with instructions for stage designers and actors or a storyboard for a film. They added written comments on how their aesthetic decisions cohere with their theoretical outlook. The examples in this section are taken from classes taught online during the COVID-19 pandemic. Developing storyboards lent itself well to the method of first working in small groups and then sharing visualizations during video conferences. In-person teaching would, of course, have given students the opportunity to test their approaches in the classroom. Either way, creative work on imagining a written dramatic text as a performance sharpened the students' sense of aesthetic options.

Several students discussed the use of particular staging techniques to demonstrate feminist empowerment. Bilodeau's *It Starts with Me*, a play with an all-female cast, addresses the ways in which climate change affects women in particular and encourages women to make their voices heard. Female performers speak lines solo or in groups, finally shouting, "Yes," "No," "It starts with me," and "Me" as they stomp their feet. Two students suggested that not having any props "reflects minimalism and contrasts [with] materialism" in "a performance" that functions "as a live consciousness-raising practice" in tune with feminist activism. Considering the work's didactic impetus, two students in the teacher training program created a director's handbook for fourteen- through eighteen-year-olds. They visually corroborated the playwright's words by imagining how they would stage the process of metaphorically and literally "seeing the big picture": "the actresses are not only going to speak but are also going to build a big picture of the Earth with puzzle pieces."

Another student team highlighted the feminist potential of Marcia Johnson's short play *Single Use*, which depicts Mitchell and Valerie's first date in a coffee shop. As it turns out, Mitchell considers serious topics off-putting and cannot fathom why Valerie would refuse a plastic straw for her milkshake and bring her own metal spoon. She indicates that she has not had any second dates yet because of her discomfiting concern with climate change and environmental destruction. Objects and sounds—like Mitchell slurping through a straw and Valerie clinking her spoon against the glass—complement the fast-paced turn-taking between a climate change ignorer and a well-informed promoter of environmentally friendly habits. Across the board, students recognized the powerful impact of visuality in theater. In *Single Use*, the visual semiotics of the coffee shop—conveyed through disposable coffee cups and plastic lids, single-serving condiment pouches, and Mitchell's fashion- and brand-conscious costume—indicate consumerism sans environmental conscience. Valerie's personal spoon, which is visually out of place, thus becomes a symbolic precursor of her opposite stance and its impact on sociability—that is, the impossibility of a second date in a social setting that thrives on throwaway materials.

Directly focusing on the use and misuse of photography in climate change communication, Paula Cizmar's short play *Appealing* addresses the danger of visually aestheticized disaster (fig. 1). Jana and J, who work for "an environmental media outlet" (Cizmar 93), are torn regarding the ethics of producing stunning photographs of catastrophic phenomena like "*a vast trash pile the size of a small island*" (94). Their conversation evolves toward rejecting "attractive" in favor of "compelling" (97) photographs. Rather than pleasing viewers' sense of aesthetic beauty or their indulgence in so-called disaster porn, Jana and J endeavor to identify images that could inspire potential readers to learn and care about anthropogenic climate change. In response to the fast-paced dialogue, which seminar participants theorized may express "our lack of time" and the necessity to grasp

Figure 1. Benedict Danner and Aline Wetzmair. Hand-drawn storyboard for Paula Cizmar's *Appealing*.

environmental issues on "a smaller, personal level," two students developed a storyboard that uses identical body postures to symbolize the characters' eventual cooperation. Projections of the debated photographs draw viewers into the characters' newly gained activist perspective.

Katie Pearl's *The Earth's Blue Heart* suggests the importance of a "relationship" (279) between humans and fish as an alternative to a merely commercial focus—a topic that the playwright derived from debates about Indigenous fishing rights. Imagining this mental journey into "the idea of [this] relationship" (279) as a film, two students developed a storyboard with detailed explanations that cover the use of colors, sounds, lighting, and camera work, culminating in a surreal state of experiencing that imagined relationship with fish instead of seeing them as a commodity. Citing multiple scholars (e.g., Baucom and Omelsky 15), they want their film to provide a psychologically meaningful emotional perspective on climate change. Repeated references to characters being or becoming "floaters" (Pearl 281) indicate those characters' evolving relationship with the blue oceanic world. As the last two panels in the students' storyboard illustrate (figs. 2 and 3), a film, as opposed to a stage performance, could easily unite within one frame both floating and standing characters

Figure 2. Michael Dietmaier and Sophie Leitner. Panel 10 of hand-drawn storyboard for Katie Pearl's *The Earth's Blue Heart*.

Figure 3. Michael Dietmaier and Sophie Leitner. Panel 11 of hand-drawn storyboard for Katie Pearl's *The Earth's Blue Heart*.

and could convey floating in ways that a stage performance could only "[i]mperceptibly" indicate (Pearl 282).

Climate change theater addresses a mind-bogglingly expansive spatial and temporal phenomenon. As in English-speaking and Western theater worlds in general, climate change artist-activists confront the challenge of inclusiveness when it comes to theater makers and audiences. Bilodeau acknowledges that CCTA's climate change activism must needs be coupled with making the voices of artists and theater practitioners of all backgrounds heard (Bilodeau, "What I Learned"), and I appreciate that the (albeit mostly English-language) CCTA corpus includes works by playwrights from all inhabited continents and of various ethnicities, genders, and sexual orientations. The study of Indigenous knowledge systems alongside non-Western theater aesthetics strengthened students' commitment to a postcolonial and transnational approach to climate change issues and their engagement through art.

The seminar participants appreciated the two-step approach of first exploring how they can apply theoretical concepts and methodologies of literary, drama, and theater studies to understanding specific works and subsequently combining such critical analysis and argumentation with the creative task of conceptualizing a stage performance or film. Developing a director's handbook or storyboard opened their eyes to details and options they otherwise would not have noticed or considered. I am optimistic that both seminars encouraged my students to do more research on climate change theater and to employ such works in their teaching.

Notes

I acknowledge with heartfelt gratitude the students I taught in the seminars described here at the University of Graz between March 2020 and February 2021. Whereas I met the first group once before the pandemic forced us to switch to online teaching, I never got to see the second group in person. Although I cannot incorporate everyone's contributions here, I appreciate the enthusiasm and hard work all students invested in the online discussions and offline assignments.

1. At the time of writing, three plays in the cycle—*Sila*, *Forward* (set in Norway), and *No More Harveys* (a one-woman play set in Alaska)—have been performed.

2. These suggestions are partially derived from one student's proposition to use Ashcroft and colleagues to apply postcolonial theory. Quotations from student work are reproduced with permission.

3. Bilodeau uses two spoken word poems by Taqralik Partridge (*Sila* 5).

4. I assigned a short journalistic text on the genesis of the term *Anthropocene* in stratigraphy (Stromberg) and a scholarly text about "planetarity" (Elias and

Moraru) as an emerging heuristic outlook in literary and cultural studies. In a class that focused on the Anthropocene, it would be worth unraveling the controversy surrounding the term as falsely privileging human agency, Western perspectives, and a nature-culture division (also see Stef Craps's contribution to this volume).

Works Cited

Aporta, Claudio. "The Trail as Home: Inuit and Their Pan-Arctic Network." *Human Ecology*, vol. 37, no. 2, 2009, pp. 131–46, https://doi.org/10.1007/s10745-009-9213-x.

Ashcroft, Bill, et al. *The Empire Writes Back: Theory and Practice in Post-colonial Literatures.* Routledge, 1989.

Baker-White, Robert. "Other Others: Dramatis Animalia in Some Alternative American Drama." *Readings in Performance and Ecology*, edited by Wendy Arons and Theresa J. May, Palgrave Macmillan, 2012, pp. 33–41.

Balestrini, Nassim W. "Anthropogenic Climate Change Condensed: Creating Community in Very Short Plays." *Journal of the Short Story in English / Les cahiers de la nouvelle*, vol. 73, 2019, pp. 121–34.

———. "Climate Change Theater and Cultural Mobility in the Arctic: Chantal Bilodeau's *Sila* (2014)." *Theater and Mobility*, special issue of *Journal of Contemporary Drama in English*, vol. 5, no. 1, 2017, pp. 70–85, https://doi.org/10.1515/jcde-2017-0006.

———. "Sounding the Arctic in Chantal Bilodeau's Climate Change Plays." *Nordic Theatre Studies*, vol. 32, no. 1, 2020, pp. 66–81, https://doi.org/10.7146/nts.v32i1.120408.

———. "Transnational and Postcolonial Perspectives on Communicating Climate Change through Theater." *Addressing the Challenges in Communicating Climate Change across Various Audiences*, edited by Walter Leal Filho et al., Springer, 2019, pp. 247–61, https://doi.org/10.1007/978-3-319-98294-6_16.

———. "'Writing Plays That *Are* Climate Change': An Interview with Chantal Bilodeau." *Journal of Contemporary Drama in English*, vol. 8, no. 1, 2020, pp. 34–46, https://doi.org/10.1515/jcde-2020-0004.

Balme, Christopher. *The Cambridge Introduction to Theatre Studies.* Cambridge UP, 2008.

Baucom, Ian, and Matthew Omelsky. "Knowledge in the Age of Climate Change." *The South Atlantic Quarterly*, vol. 16, no. 1, 2017, pp. 1–18.

Bilodeau, Chantal. *Forward.* Talonbooks, 2018.

———. Introduction. Bilodeau and Peterson, pp. 15–21.

———. *It Starts with Me.* Bilodeau and Peterson, pp. 67–72.

———. *No More Harveys.* Talonbooks, 2023.

———. *Sila.* Talonbooks, 2015.

———. "What I Learned about Gender Parity and Racial Diversity from Running a Global Participatory Initiative." *Artists and Climate Change*, 18 Apr. 2019, artistsandclimatechange.com/2019/04/18/what-i-learned-about-gender-parity-and-racial-diversity-from-running-a-global-participatory-initiative/.

———, editor. *Where Is the Hope? An Anthology of Short Climate Change Plays.* Climate Change Theatre Action, 2018.

Bilodeau, Chantal, and Thomas Peterson, editors. *Lighting the Way: An Anthology of Short Plays about the Climate Crisis.* Centre for Sustainable Practice in the Arts / The Arctic Cycle, 2020.

Chaudhuri, Una. "The Silence of the Polar Bears: Performing (Climate) Change in the Theater of Species." *Readings in Performance and Ecology,* edited by Wendy Arons and Theresa J. May, Palgrave Macmillan, 2012, pp. 45–57.

Cilano, Cara, and Elizabeth DeLoughrey. "Against Authenticity: Global Knowledges and Postcolonial Ecocriticism." *ISLE: Interdisciplinary Studies in Literature and Environment,* vol. 14, no. 1, 2007, pp. 71–87, https://doi.org/10.1093/isle/14.1.71.

Cizmar, Paula. *Appealing.* Bilodeau and Peterson, pp. 93–98.

Elias, Amy J., and Christian Moraru. "Introduction: The Planetary Condition." *The Planetary Turn: Relationality and Geoaesthetics in the Twenty-First Century,* edited by Elias and Moraru, Northwestern UP, 2015, pp. xi–xxxvii.

Gaard, Greta. "Where Is Feminism in the Environmental Humanities?" *Environmental Humanities: Voices from the Anthropocene,* edited by Serpil Oppermann and Serenella Iovino, Rowman and Littlefield, 2017, pp. 81–95.

Hutcheon, Linda. "Eruptions of Postmodernity: The Postcolonial and the Ecocritical (1993)." *Greening the Maple: Canadian Ecocriticism in Context,* edited by Ella Soper and Nicholas Bradley, U of Calgary P, 2013, pp. 123–44.

Johnson, Marcia. *Single Use.* Bilodeau, *Where Is the Hope?,* pp. 153–56.

May, Theresa J. *Earth Matters on Stage: Ecology and Environment in American Theater.* Routledge, 2020.

Pearl, Katie. *The Earth's Blue Heart.* Bilodeau, *Where Is the Hope?,* pp. 279–84.

Rajewsky, Irina. "Intermediality, Intertextuality, and Remediation: A Literary Perspective on Intermediality." *Intermédialités/Intermediality,* vol. 6, fall 2005, pp. 43–64, https://doi.org/10.7202/1005505ar.

Stromberg, Joseph. "What Is the Anthropocene and Are We in It?" *Smithsonian Magazine,* Jan. 2013, www.smithsonianmag.com/science-nature/what-is-the-anthropocene-and-are-we-in-it-164801414/.

Zapf, Hubert. "Cultural Ecology of Literature—Literature as Cultural Ecology." *Handbook of Ecocriticism and Cultural Ecology,* edited by Zapf, De Gruyter, 2016, pp. 135–53.

Parker Krieg

Climate Change Narratives, Publics, and the Professional-Managerial Class

This essay discusses a recent seminar, titled Climate Change Narratives: Professionals and Publics, at the University of Helsinki. The seminar brought together graduate and advanced undergraduate students in English and North American studies with students from the university's graduate program in environmental change and global sustainability. The course takes up the challenge of a multidisciplinary classroom by exploring the division of labor in climate literature—that is, through a sociology of professionals, intellectuals, and experts, and in relation to theories of publics and counterpublics. This approach gives students and teachers a critical lens to situate themselves alongside these narratives as would-be professionals within contemporary economies of knowledge. Furthermore, it asks how climate change may be transforming these relationships, whether by exacerbating the dissolution of trust in neoliberal societies or by inspiring citizen science and new political organization.

A key dimension of climate literature, which includes the scholarly criticism that produces literature as an object, is the social division of class, labor, and education. While thinkers like C. Wright Mills, Andre Gorz, Eric Olin Wright, Thorstein Veblen, and Alvin Gouldner have analyzed various aspects of middle-class politics, it is Barbara Ehrenreich and John

Ehrenreich who are most known for their formulation of the professional-managerial class (PMC). The PMC has historically included scientists, lawyers, doctors, engineers, and teachers but may be updated to include nonprofit consultants and people in communicative media jobs. As a class whose interests lie "between labor and capital," the PMC exhibits an ambivalent cultural politics (Ehrenreich and Ehrenreich, "Professional-Managerial Class" 17). Subservient to those above them yet feeling a duty to enact social reform, this debated new class has been criticized by the left for hypocritical careerism and by the right for elitism. Often these critiques focus on the conspicuous consumption in the cultural politics of middle-class liberals. When discussions of the PMC reach the mainstream, they often fall prey to the same pettiness that the critics themselves wish to diagnose. Because climate change and environmentalism invoke appeals to both scientific authority and personal habits, reactions are strong.

The nationalist reaction that directs its ire toward climate and migration often takes aim at what it sees as out-of-touch domestic elites. Yet the resurgence of PMC discourse in recent years has underscored that class's diminishing economic status and loss of influence, leading to a new workplace militancy in the education and media sectors (Ehrenreich and Ehrenreich, *Death*). Here, both the resentment of national conservatives and the militancy of the educated precariat can be understood as effects of neoliberal globalization whose hold on subjectivities and the state is a constitutive part of the climate emergency. In this conjuncture, professionals are not merely an object of public discourse; they are part of the public, and they are likewise positioned to redefine the relationship of knowledge to society.

Focusing on the professional class and publics in climate literature brings the environmental humanities together with the sociological wing of cultural studies. "One of every five employed Americans is a professional," writes Jeffrey J. Williams, who makes his case for "teaching the professions" as a theme in the literature classroom: "If we teach in a college or university, most of the students in front of us are there to become professionals and managers" (69). Ignoring this fact when it comes to the environmental humanities would do a disservice to students and the social challenges they will face in their prospective careers. How will students put their values into action? The last century witnessed a shift from the "social-trustee professional" to the "expert" for whom knowledge is an entrepreneurial property (Brint 11). Each bears a different relationship (real and imagined) to the public. Feminist critics have challenged the male-dominated institutions of the former and the mercenary attitude of the latter by employing "strong objectivity" in community-oriented research (Harding). In literature, Richard Ohmann argues that the concerns of the professional middle class are central to the formation of the postwar literary canon ("Shaping" 209). Perhaps no one has devoted more attention to what it means to be a member of the PMC as a critical professional in the university. In his analysis of the reproductive role the PMC plays in capitalism, Ohmann nevertheless credits it with accomplishments such as "making food and water safer, offering various social services, . . . [and] help[ing] organize an embryonic welfare state" (*Politics* 93). As this group's status is so closely linked with the university, it is important to understand its past, present, and possible futures when teaching environmental texts.

Both creative and nonfiction writers are in a sense disavowed professionals with varying degrees of autonomy from the social arrangements they depict. Writing for the *New-York Tribune* in 1854, Karl Marx championed Charlotte Brontë, Charles Dickens, and Elizabeth Gaskell, who "have issued to the world more political and social truths than have been uttered by all the professional politicians, publicists, and moralists put together." These novelists depicted the middle class as "servile to those above, and tyrannical to those beneath them," and novelists continue to imaginatively influence publics and counterpublics today. This social perspective stands in contrast to contemporary approaches in ecocriticism that seek to separate the aesthetic experience of ecology from instrumental knowledge by way of sublime bewilderment or ontological decentering. In his multivolume *Social History of Art*, Arnold Hauser, a student of Georg Lukacs, attributes the emergence of this kind of Romanticism in nineteenth-century Germany to the powerlessness of its middle class, who proposed private "spiritualizing" solutions to social contradictions. "As a form of emotionalism," he writes, "the romantic movement still had a direct link with the revolutionary tendencies at work in the middle class[;] as a form of idealism and supernaturalism, on the other hand, it became increasingly remote from progressive middle-class thought" (113). Sean McCann and Michael Szalay identify a similar turn in cultural politics following the 1960s New Left to what they call "magical thinking," in that literature and culture come to valorize ecstatic experience as a way of doing politics through a rhetoric of being (e.g., symbolic consumption) rather than the more difficult task of building power through labor and party organization (452). This tendency can be understood as a symptom of a professional middle class that is torn between its sociotechnical role and a desire for creative autonomy, enjoying cultural influence but lacking the political organization to transform society.

The seminar presented an opportunity for both critical and reconstructive readings of professionals and publics by asking how each can reframe contemporary climate narratives. For example, the focus on professionals encourages students to observe how problems of knowledge or aesthetics in a narrative might express or engage an underlying sociopolitical antagonism beyond the intentions of characters. Students might consider how professional training shapes the questions and expectations they bring to the text. This practice allows students to make distinctions between primary and secondary knowledge and to consider how this distinction operates at the level of narrative form. Conversely, they might explore how climate and environmental emergencies become depoliticized through appeals to a lack of knowledge. Reading this way, students may come to recognize how implicit interests are made transparent or complicated by the text. Concurrently, reading for publics invites students to question how narratives represent various formations of a public sphere, to ask who is included and excluded, and to reflect on how readers are enlisted or otherwise tasked with participating in a public by the text. Lastly, this framing asks students to consider how climate change necessitates citizen-science initiatives and new ways of producing knowledge in society. What follows is an overview of these three dimensions, illustrated by the novels and topics addressed by the seminar participants.

Experts and Publics

Our first unit explored the expert-public relationship through Nathaniel Rich's *Odds against Tomorrow*, which famously anticipated the flooding of New York City by Hurricane Sandy. The novel dramatizes the professionalization of speculation as corporate disaster insurance, facilitated by the main character's hyperactive imagination. Importantly, it thematizes the way that speculation itself, a favored mode of thinking within environmental criticism, is subsumed into high finance as secular prophecy. Our discussions were supported by secondary readings that contextualize the rise of scenario planning in Cold War think tanks as a practice of managing perpetual instability. Following the oil shocks of the 1970s, "turbulent world" models influenced the speculative turn in financial markets and the expansion of the United States military (Cooper 170). Assigning *Odds against Tomorrow* provides an opportunity to discuss multiple uses of environmental specula-tion set against the backdrop of New York City's finance, insurance, and real estate economy. At the same time, the novel gives students a vehicle to discuss the existential stakes of environmental risk while exposing the cycles of community experimentation and corporate change-making.

Our discussions considered the relationships and attitudes that each character has toward risk in their personal and work lives. This allowed students to analyze different relationships to knowledge. Drawing on a supple-mentary text by Nico Stehr and Reiner Grundmann, two students from the natural sciences observed that the main character, rather than producing original information, is a second-order observer. He creates stories out of existing information and extrapolates to the point of uncertainty, which he deploys to induce emotional states in potential clients. By instrumentalizing the kind of speculation highlighted above, the protagonist turns hypothet-ical climate knowledge into capital for his employer. The public is less at-tracted by this corporate insurance game than by the character's cultural status as prophet of the flood. They believe he has answers; he has none. This reading challenged the tendency of students to think of professionals and publics separately rather than as related parts of a social process.

Next, we considered how uneven risk in the built environment is nar-rated. Drawing on readings from Jane Bennett and Bonnie Honig, discus-sion turned to infrastructures in disrepair owing to decades of privatization. Among the novel's ruined objects of post-flood Manhattan lie the corpses of those who were unable to escape. The free indirect description of the subway scene analyzes the cognitive behavioral response of the victims. Psy-chologizing the storm at this scale shifts the blame to the victims for mak-ing the "wrong" decision in a moment of crisis. It evades the structural causes of disproportionate risk exposure, chief among them the heating of the atmosphere. One student, observing the danger posed by the debris, suggested that Bennett's new materialism can defamiliarize human agency as a guide to assessing environmental vulnerability at multiple scales. An-other suggested that flattening human and nonhuman agency reinforces a managerial mindset and masks the social power of the expert. We left this generative tension unresolved to inform future discussions.

Publics and Resilience

The second unit explored publics and resilience through *Archipelago*, by the Trinidadian British author Monique Roffey, which centers a middle-class, multiracial family that is traumatized by a hurricane. Their journey traces an environmental memory of colonial and anthropogenic change across the Caribbean. This postcolonial context challenges essentialist notions about race and culture in disaster response and ecotourism. Of particular interest are the novel's mediascapes, which elicit sympathetic identification of the characters with landscapes and people beyond their immediate surroundings. Class discussions focused on passages where characters recognize themselves in images of disaster footage from other Caribbean and Latin American countries. In spite of the dominance of Global North media conglomerates, this recognition becomes one source of a public within the novel. A much larger public is forged out of parallel socioecological histories, catastrophic presents, and a more-than-human ethos of shared vulnerability.

Michael Warner's essay "Publics and Counterpublics" helped us delineate the different kinds of publics in climate narratives. Asking which social imaginaries are appealed to by (and in) texts, Warner reminds us that "a public is a space of discourse organized by nothing other than discourse itself." While this "autotelic" conception of the public (50) seemingly excludes material environments as potential vectors of affiliation, our readings of *Archipelago* challenged this. In short responses, it became clear that students did not find an easy identity between the novel's understanding of its public and simplistic notions of community. It can therefore be argued that the novel challenges the opposition that holds the so-called rootless modern West against the implied localism and communal bonds of the non-West. Rather, the publics in *Archipelago*'s locations are created out of geohistorical continuities, the private imaginations of characters, and the mediated confrontation with divisions of class, race, and environmental risk. In this light, one student described the neo-imperial tourist industry as a counterpublic that is revealed as the characters compose an altogether more layered public through their travels. Attending to this complexity is necessary in constructing the "relation among strangers" needed for climate solidarity (Warner 55). Such solidarity highlights the work of novels and media in building shared horizons of experience.

Citizen Science and Knowledge Workers

Our final unit explored relations between publics and scientists through Barbara Kingsolver's novel *Flight Behavior*. Kingsolver's well-known work of cli-fi addresses a wayward butterfly migration in a narrative of rural class mobility made possible by college education and divorce. The transformation in the protagonist's life occurs after her collaboration with researchers from more cosmopolitan settings. Our discussions focused on the novel's portrayal of local contributors to the project, in particular the gendered dimensions that persist between local knowledge and expertise. Here too, the relationship between "knowledge work and the commons" that Heather Houser identifies in Kingsolver provided additional opportunities to connect the work of environmental researchers to changes in the post-Fordist economy (104). These connections raised the political implications of citizen-science projects like the one in *Flight Behavior.*

Students responded to the novel's critique of the assumptions the scientists make about the environmental habits of the locals and shared similar examples from their own respective regions and countries. One student wrote about how the novel changed the way she viewed her own major and the asymmetrical privilege that exists between her university and the surrounding community in the American Southwest. Others focused on the portrayal of social constraints on scientists. For researchers, depression felt at the loss of ecosystems is compounded by the disillusionment of discovering that their work in the so-called knowledge economy often only has value if it serves corporate or state interests. Nevertheless, students were ultimately drawn to the optimism of multidirectional solidarity between scientists and community members.

This unit offers the most potential for teachers to incorporate external activities and assignments related to citizen science. This could involve a growing array of apps, participation in local initiatives, service-learning, or other kinds of community engagement to create opportunities for additional reflection on the social nature of knowledge. Scheduling constraints permitted only a narrative exploration of citizen science in this seminar. However, by encouraging students from multiple majors and disciplines to bring their own fields into the classroom, we created a forum that places a critical and sociological understanding of professional (and public) life squarely at the heart of the environmental humanities. If, as Stephanie LeMenager suggests, the struggle within cli-fi is a "struggle for genres" capable of making the social world of climate legible in new ways, a lens that focuses on professional and managerial divisions of labor may reveal a good deal about the social terrains that overdetermine climate politics.

For a field experience, I planned a trip to Helsinki's Museum of Contemporary Art Kiasma to view *Weather Report: Forecasting the Future,* an exhibit by Finnish artists from the Venice Biennale. Kiasma had recently hosted a curator talk by T. J. Demos on the aesthetics of climate emergency along with exhibits by another Finnish organization, Bioart Society, on the theme of coexistence. The goal of the visit was to explore different ways of thinking about environmental publics in public. Such field experiences, however, were suspended by COVID-19. As the course moved online, discussion forums took the place of reading journals, and students drew striking connections between climate change and COVID-19. They shared existential concerns about public health, science communication, and international politics. Building on short unit essays, they developed final research papers that investigated the relationship of professionals and publics in their own disciplinary fields using what we had learned from climate narratives as a heuristic lens. The goal of the seminar is to enable a double reflexivity and to position one's reading sociologically rather than relying on an imagined relationship to society or an imagined relationship to a historical image of nature. Better still, it may prepare students for the actual relationships they will face upon entering the workforce and help them decide what kind of professional and public they want to be on a planet that needs both.

Works Cited

Bennett, Jane. "The Force of Things: Steps toward an Ecology of Matter." *Political Theory*, vol. 32, no. 3, 2004, pp. 347–72.

Brint, Steven. *In an Age of Experts: The Changing Role of Professionals in Politics and Public Life*. Princeton UP, 1994.

Cooper, Melinda. "Turbulent Worlds: Financial Markets and Environmental Crisis." *Theory, Culture and Society*, vol. 27, nos. 2–3, 2010, pp. 167–90.

Ehrenreich, Barbara, and John Ehrenreich. *Death of a Yuppie Dream: The Rise and Fall of the Professional-Managerial Class*. Rosa Luxemburg Stiftung, 2013, rosalux.de/fileadmin/rls_uploads/pdfs/sonst_publikationen/ehrenreich_death_of_a_yuppie_dream90.pdf.

———. "The Professional-Managerial Class." *Radical America*, vol. 11, no. 2, 1977, pp. 7–31.

Harding, Sandra. *Objectivity and Diversity: Another Logic of Scientific Research*. U of Chicago P, 2015.

Hauser, Arnold. *A Social History of Art*. Vol. 3, Vintage, 1958.

Honig, Bonnie. "The Politics of Public Things: Neoliberalism and the Routine of Privatization." *No Foundations: Interdisciplinary Journal of Law and Justice*, vol. 10, 2013, pp. 59–76.

Houser, Heather. "Knowledge Work and the Commons in Barbara Kingsolver's and Ann Pancake's Appalachia." *MFS: Modern Fiction Studies*, vol. 65, no. 1, 2017, pp. 95–115.

Kingsolver, Barbara. *Flight Behavior.* HarperCollins Publishers, 2012.

LeMenager, Stephanie. "Cli-Fi and the Struggle for Genre." *Anthropocene Reading: Literary History in Geologic Times,* edited by Tobias Menely and Jesse Oak Taylor, Penn State UP, 2017, pp. 220–38.

Marx, Karl. "The English Middle Class." *New-York Tribune,* 1 Aug. 1854, marxengels.public-archive.net/en/ME1912en.html.

McCann, Sean, and Michael Szalay. "Do You Believe in Magic? Literary Thinking after the New Left." *The Yale Journal of Criticism,* vol. 18, no. 2, 2005, pp. 435–68.

Ohmann, Richard. *Politics of Knowledge: The Commercialization of the University, the Professions, and Print Culture.* Wesleyan UP, 2003.

———. "The Shaping of a Canon: U.S. Fiction, 1960–1975." *Critical Inquiry,* vol. 10, 1983, pp. 199–223.

Rich, Nathaniel. *Odds against Tomorrow.* Picador, 2013.

Roffey, Monique. *Archipelago.* Penguin, 2012.

Stehr, Nico, and Reiner Grundmann. *Experts: The Knowledge and Power of Expertise.* Routledge, 2011.

Warner, Michael. "Publics and Counterpublics." *Public Culture,* vol. 14, no. 1, 2002, pp. 49–90.

Williams, Jeffrey J. "Teaching the Professions." *Radical Teacher,* vol. 99, no. 4, 2014, pp. 69–75.

Thomas Hallock

Words in the World:
The Work of an Environmental Literature Course in a Coastal Florida City

Given the scale of the crisis, a challenge for teaching climate change lies in localization. How can we break down a problem this big? How do we foster concern without sounding the apocalyptic alarms? I teach at an urban campus on the Gulf Coast of Florida, adjoining Tampa Bay, where the consequences of vanishing glaciers and ice shelves literally wash against my workplace. Global warming and sea-level rise, the United Nations reports, are certain to result in increased "tropical cyclone winds and rainfall" (Pörtner et al. A3); that means, each fall for the remainder of my career, I must plan syllabi with a week off for hurricanes. Coastal communities, the UN adds, "face challenging choices" that must balance "costs, benefits and trade-offs" (C3); these decisions, made at City Hall, come from policymakers whom I know on a first-name basis and who have visited my classes. Global warming (if left unchecked) will result in sea levels rising between ten and thirty feet over the next century (Strauss), which means that for the children's children of my current students, our campus and university—really the entire state—will be underwater. So how do we shrink the undeniable need for stewardship to fit a sixteen-week syllabus?

I start by grounding courses on nature writing, which I have taught almost every spring for the past fifteen years, in our immediate environs.

I focus in particular on the creeks and waterways closest to our campus. By drawing attention to city nature, I steer students away from many of the hackneyed tropes of the genre (wilderness as spiritual retreat, nature writer as self-righteous loner, green jeremiads) and toward a more nuanced engagement with the everyday. (The field trips are also a welcome source of images for campus promotional materials.) These courses, geared to upper-level English elective and master's degree students, follow a three-part plan. Using an old chestnut from Robert Scholes, that English study is about the production and reception of texts in a social setting (30), I ask students to work through this triangular model: direct observation, or field experience (the social setting); reading and research, theoretical and literary, in the environmental humanities (reception); and attention to the writing process, from journaling through revision to a final essay (production). This approach works best with advanced students, who readily grasp that all three elements—field trips, theory, deep revision—are essential. Cut out one part, I tell my class every semester, and the triangular structure will collapse.

Reading and writing assignments reinforce the importance of each side. Drawing from my background in eighteenth-century studies, I provide students with rhetorical models from three "lost" forms—a botanic essay, an ornithological biography, and a "ramble"—one of which they will adopt for an original piece of writing. In studying and imitating the examples from their assigned readings, students learn the basics of birding and plant collection, and over the years our classes have assembled an ongoing "cabinet of curiosity" that now provides me with models. Alongside neoclassical authors such as William Bartram, John James Audubon, and Erasmus Darwin, we read contemporary reflections by Hope Jahren, Susan Orlean, and others. After students have composed their own field-based essays, the class then turns to a more theorized question: How do we find nature in the city? The students work through Greg Garrard's *Ecocriticism*, a breakneck romp through a centuries-old tradition, which divides environmental thought into several categories. Garrard offers chapters on the pastoral, the georgic, animals, the catastrophic, and so on. But what language do we use to describe nature in the city? Lacking terms for our place, we then explore other options—drawing from shorter articles by bell hooks, spatial theorists like Henri Lefebvre and Michel de Certeau, the landscape architect Anne Whiston Spirn, and Jenny Price. At the same time we walk the streets and check out nearby waterways, connecting local histories with firsthand observation. A second writing assignment theorizes these observations of the everyday. Finally, a third assignment asks students to synthesize, braiding together the different strands of the course into one narrative reflection. This capstone project necessitates a more process-oriented approach, and as we work through the last (and more challenging) exercise, we turn to writing guides such as Anne La-mott's *Bird by Bird*, which leavens the challenges of self-creation with story, and Roy Peter Clark's *Writing Tools*, which provides students who are close to graduation (and rightly anxious to sharpen their skills for the workplace) with a "finish" to their prose.

What I most like about this curriculum is the scale. The three-part pedagogical structure makes the overwhelming (climate change) approach-able. It gives students the framework they need to engage on their terms without falling back on tired formulas. The outdoor component prompts individual buy-in; students take on the challenge of a nature essay because, with the proper field experience, a subject will announce itself to them. One of my favorite essays in recent years is a reflective piece, "Sewing Box," by a graduate student named Hannah Gorski. This essay describes the re-covery of an old upholstered sewing box that Gorski found on one of our class canoe jaunts.

The essay is a model of close and careful description, opening to much broader questions. As she notes, Gorski did not set out to find her titular sewing box. On the day she describes, a small group of us were looking for a rotting great blue heron that our class had seen on the creek during a previous canoe paddle. Tangled in monofilament, the heron had drowned, and our thought was to preserve it, with the guidance of a field naturalist, as an ornithological specimen.[1]

The assignments were already building upon one another. Gorski was following up on her first essay, the ornithological biography. The natural-ist we had consulted had given us detailed instructions for the bird's pres-ervation: We should retrieve the carcass with a plastic box, wrap the re-mains in window screen and staple it shut, and then toss the entire thing over a fire ant mound. After a week, once the smell cleared, we should have feathers and bones. That Gorski had even agreed to pull a rotten long-legged wader from the creek indicates that the course was on track. She had latched onto the three-part shape of environmental writing—engagement with experience, precedent, and writing craft. We set out in search of the filthy heron because Gorski rightly sensed a story; to accumulate the detail that drives a nature essay, one must take to the field. She was also sifting through theoretical paradigms. In this case, Gorski would draw from

Spirn, whom she quotes: "To read landscape is also to envision, choose and shape the future: to see, for example, the connections between buried, severed stream, vacant lands and polluted river, and to imagine rebuilding a community while purifying its water" (399).

"Purifying" served as the cue. Mercifully, we did not find the heron that day. (The carcass had presumably broken from the monofilament and floated off to sea.) As a consolation prize, Gorski retrieved a beaten-up sewing box. This artifact, with cracked white upholstery and red velvet lining, clearly once had been loved. Gorski took home the filthy object, and in her final reflective essay, she described the act of cleaning as an active metonymy for the polluted tidal creek:

> Water from the garden hose dilutes Salt Creek from the fibers of the crimson cloth. Spools of thread, the bobbins, and all the buttons lighten as the water rinses each surface. Loose string intensified the gross into a spider web of leaves, twigs, buttons, card-like paper, and on each object, I know, an unknown pollutant lingers. Leaves drop from hidden spaces. I can only imagine what has sifted through the boundaries of this container. . . . I place all the items in a cooler and carry the box and trinkets upstairs to let them dry near an open window. I'm not sure if I regret my decision to clean the box. My boyfriend says the box belongs in the dumpster. I tell him, "It's for school." (43)

The step-by-step process provides the author with a stockpile of detail. Descriptions are concrete, and the author's voice modulates effectively between self-deprecation and sincerity.

Drafts provide a backstory to this composition. Taking Lamott's helpful guide to heart, Gorski titled her first effort "Shitty First Draft." She describes the exchange with her boyfriend, which anchors the final essay:

> I have paddled Salt Creek twice before. The second time, my professor salvaged a sewing box near the 4th Street Bridge, half submerged in the water. Its scarlet fabric flashed from the gritty aquatic. I volunteered to restore the box. I was curious. I wrapped it in a trash bag and drove it home. When I told my boyfriend, he said the recycle bin would appreciate the box. I told him, "No, it's for school."
>
> I carried the box to the patio and rinsed it off with the hose. I rinsed the spools of thread, the bobbins, and all the buttons. I tipped the box to drain the water. I could smell the creek—it's a dirty and sad stench. Nobody sees this because the creek is blocked from view. I didn't like touching the objects and I should have worn gloves, but I washed my hands over and over. I placed all the items in a cooler and carried the

box and trinkets upstairs to my apartment to let them dry near an open window.

The draft shows how an author slowly worked toward her story. The more flowery phrasing, edited out, indicates an earlier struggle with tone ("scarlet fabric flashed from the gritty aquatic"). In a supplemental draft, Gorski even toys with the idea of a "bird's eye" view of the creek. A page of fragments captures this and other false starts:

> His pupil is spoiled now. . . . I see him concentrating on his last staggered breaths, unable to clean the polluted creek from his view.

> Salt Creek, this wasted waterway, is not my ideal paddle. Mangroves crowd the passage, their. . . .

> I remember the mangrove roots snaring the generic goods. I expected empty potato chip bags. I expected sports drink bottles and beer cans. I did not expect that blue, tangled mass.

The paragraphs stop mid paragraph, even mid sentence, as the author strains to link story and concept. Her descriptions gain their shape, however, after a review of Spirn on Philadelphia's Mill Creek. In a second fragment, titled "Spirn Draft," Gorski writes, "Salt Creek's condition and the state of South St. Pete intersect in a way not uncommon for many communities. I read Anne Whiston Spirn's essay about a sidelined neighborhood in West Philadelphia developing on top of a drained watershed. . . ." Once Gorski establishes this theoretical connection, the physical object acquires broader relevance.

It took me almost twenty years to develop this curriculum and to unlearn what we so often mean by "nature" in an environmental humanities class. I first came to my campus as a visiting instructor in 2001, when I led an experimental seminar called Rivers of Florida. This course worked on an innovative premise: that we read about Florida's rivers from every angle possible, then go for a paddle. Some of those angles, however, demonstrably put people off. Well-intentioned readings from the popular press (Brunais; Hiaasen; Funk) failed to nurture a connection to the environment that students could feel. Nature is not church. Nature writing that sounds like a sermon will never serve its rhetorical purpose. And, even as the consequences of sea-level rise become increasingly apparent, ranting about the end of the world will get us nowhere.

When I returned to campus a few years later, this time on a tenure track, I organized a new version of the course around the Hillsborough, a lovely blackwater river that runs from its headwaters in the Green Swamp Wildlife Management Area through downtown Tampa, into the bay, and finally into the Gulf of Mexico. From 2008 to 2011, I used the course title Rivers: From the Swamp to the Bay, and we followed what seemed to be a "natural" progression—swamp to urbanized channel, wilderness to city. I observed, however, that the enrollment population self-selected (mostly white and middle-class, already comfortable outdoors or in a sailboat). Talented writers from outside this group reported fearing that their lack of experience in wild settings would compromise their grades. In an end-of-semester assessment, for example, one noted that he had hoped to go back to the river to shore up details for concrete focus but had felt "too chicken-shit to canoe by myself." Again and again, I have found that in my attempts to engage outdoor settings, many student chafe against the conventions of "nature."

A focus on everyday life, by contrast, can break down the clichés. Starting a course with observation, cast in a pre-Romantic mold, helps students limn the distinction between the built and natural environment and, as in Gorski's case, even turn the natural history model inside out. Our discussions of the writing process, built into the syllabus, provided Gorski and her classmates with a method for framing more complex engagement with our immediate setting. Avoiding the shortcuts typically afforded a nature essay (literally the bird's-eye view), Gorski worked through the tired tropes. The theoretical readings helped, as the setting of a city creek required some reframing. Writing from her terms, with a concern that did not require explanation, she cultivated an air of sympathy that ultimately renders politics moot. Our current crisis does not require grandstanding or indoctrination. What we need is a pedagogy that is theorized yet immediate, with real classroom strategies that help us see where we are—and where future dangers lie.

Note

1. For such excursions, students on my campus rent equipment from the outdoor recreation department, which requires that they sign a liability waiver and, in many cases, work with a trained guide. I am far more cautious in pedagogical situations than in personal settings. In my experience, campus recreation staff members have been delighted to collaborate on academic activities, which for me greatly lessens concerns about safety and liability.

Works Cited

Audubon, John James. *Ornithological Biography; or, An Account of the Habits of the Birds of the United States of America.* Edinburgh, 1831–39. *Biodiversity Heritage Library,* www.biodiversitylibrary.org/bibliography/48976.

Bartram, William. "Anecdotes of an American Crow." *Travels and Other Writings,* by Bartram, Library of America, 1996, pp. 573–76.

Brunais, Andrea. "The River's Health—and a Happy Ending." *La Gaceta,* 23 Nov. 2007, p. 13.

Certeau, Michel de. *The Practice of Everyday Life.* U California P, 1984.

Clark, Roy Peter. *Writing Tools: Fifty Essential Strategies for Every Writer.* Little, Brown, 2006.

Darwin, Erasmus. *The Botanic Garden: A Poem in Two Parts.* London, 1799. *HathiTrust,* babel.hathitrust.org/cgi/pt?id=hvd.32044107257677.

Funk, Ben. "Man vs. Nature: Day of Reckoning." *St. Petersburg Times,* 18 July 1971, p. B1.

Garrard, Greg. *Ecocriticism (The New Critical Idiom).* Routledge, 2004.

Gorski, Hannah. "Sewing Box." *Salt Creek Journal: Nature, Community, and Place in South St. Petersburg,* edited by Gorski et al., Tampa Bay Writers Network, 2016, pp. 41–35.

Hiaasen, Carl. "If You Like Polluted Rivers, You'll Love This." *St. Petersburg Times.*

hooks, bell. "homeplace (a site of resistance)." *Yearning: Race, Gender, and Cultural Politics,* by hooks, South End Press, 1990, pp. 41–49.

Jahren, Hope. *Lab Girl.* Knopf, 2016.

Lamott, Anne. *Bird by Bird: Some Instructions on Writing and Life.* Bantam Doubleday Dell Publishing Group, 1980.

Lefebvre, Henri. *The Production of Space.* Blackwell, 1991.

Orlean, Susan. *The Orchid Thief.* Random House, 1998.

Pörtner, H.-O., et al., editors. "Summary for Policymakers." *Special Report on the Ocean and Cryosphere in a Changing Climate,* Intergovernmental Panel on Climate Change, 2019, ipcc.ch/srocc/.

Price, Jenny. "Thirteen Ways of Seeing Nature in LA: Part I." *The Believer,* no. 33, 1 Apr. 2006, www.thebeliever.net/thirteen-ways-of-seeing-nature-in-la/.

Scholes, Robert. *Textual Power: Literary Theory and the Teaching of English.* Yale UP, 1985.

Spirn, Anne Whiston. "Restoring Mill Creek: Landscape Literacy, Environmental Justice and City Planning and Design." *Landscape Research,* vol. 30, no. 3, July 2005, pp. 395–413.

Strauss, Benjamin H., et al. "Carbon Choices Determine U.S. Cities Committed to Futures below Sea Level." *Proceedings of the National Academy of Sciences,* vol. 112, no. 44, 12 Oct. 2015, www.pnas.org/content/112/44/13508.

Part III

Texts

Magdalena Mączyńska

Attention, Connection, Dialogue: Teaching Barbara Kingsolver's *Flight Behavior* in the Climate Fiction Classroom

When I teach climate fiction to undergraduates—whether at a liberal arts college or at a maximum security prison—I start by asking myself what goals should be guiding my work. Is it my role to introduce students to the basic science of climate change? (Indeed, I often find myself in the absurd position of a literature professor explaining atmospheric chemistry.) Or is my first duty to emphasize the social justice implications of the climate crisis? Is it to foster activism? to address climate anxieties? or to field urgent questions: "Is it too late?" and "What can I do to help?" Where among these pressing matters do I find room for the slow, meticulous work of literary analysis? One text that helps me bring together these diverse pedagogical goals is Barbara Kingsolver's *Flight Behavior*. The novel follows an Appalachian housewife, Dellarobia Turnbow, as she learns about climate change from Ovid Byron, a Black ecologist specializing in monarch butterflies. The plot of Kingsolver's eco-bildungsroman unfolds against a background of socioeconomic precarity and global migration. The narrative tackles the complexities of climate activism, denialism, and media representation. As a fictional introduction to the white American experience of climate change under late capitalism, *Flight Behavior* has few peers.

More importantly, the work of *Flight Behavior* goes beyond thematic engagement with the climate crisis. Kingsolver not only registers the range of climate impacts on human and nonhuman life but also questions the conceptual structures underpinning modern consumer culture: ecological inattentiveness, hyperindividualism, and what the ecophilosopher Val Plumwood calls "monological" (that is, one-directional, inimical to mutual adaptation) relationships with the more-than-human world (63). Through illuminating and resisting these structures, the novel invites readers to rethink the way they conceptualize "nature" and their place within it.

My teaching of *Flight Behavior* is informed by three guiding principles, which also serve as unofficial learning goals for my climate fiction courses: epistemological attentiveness, ecosystemic awareness, and dialogic communication. In the sections that follow, I offer examples of how Kingsolver's narrative promotes each goal. These examples anchor the way I teach the novel and serve as foci for class discussions, assignments, and group activities. I also point to writers, such as Plumwood, Rob Nixon, Kathryn Yusoff, and Donna Haraway, whose theoretical work informs my reading of Kingsolver. Although I assign theory only in more advanced classes, I find that students at all levels can benefit from the strategic introduction of key theoretical concepts. Finally, I offer suggestions for classroom activities that invite students to begin practicing attentive observation, relational thinking, and dialogic engagement.

Epistemological Attentiveness

From the opening scene, in which the protagonist hikes up a forested mountain for an adulterous tryst, only to be stopped by a colony of monarch butterflies, Kingsolver's novel foregrounds the importance of perception. Early on in her ascent, Dellarobia looks down at the house in which she has spent her married life and is shocked by the realization that she had never seen her home from this vantage point: "She only looked out of these windows, never into them" (3). The subtle shift in perspective initiates an avalanche of epistemological shifts that will reshape the protagonist's understanding of the world. When Dellarobia is confronted with the unfamiliar sight of wintering butterflies, she lacks the conceptual and visual apparatus (vanity had prevented her from putting on eyeglasses) to understand what she is seeing. She frantically scans her mind for possibilities: a hornets' nest, a swarm of bees, an armadillo, dead leaves, pine cones, grape clusters, a fungus, and, finally—as sunlight hits the insects' orange wings—flames (12–13). In the end, she interprets the "fire" as a divine vision directed at her planned infidelity. Only on her second visit to the site, this time wearing glasses, does Dellarobia perceive the monarchs and reflect on her error: "How had she failed to see them? . . . She'd been willing to take in the run of emotions that stood up the hairs on her neck, the wonder, but had shuttered her eyes and looked without seeing" (52). The novel returns to the same spot several more times—most notably in the apocalyptic finale—to mark the protagonist's perceptual evolution.

My teaching of *Flight Behavior* begins with a close analysis of the opening scene and its epistemological concerns. I ask students to sketch Dellarobia's field of vision on a piece of paper to encourage close attention to the text's rich descriptive layers. Like the protagonist, we continue returning to this primal site throughout the semester to track its evolving meanings. Close reading, a time-honored staple of the literature classroom (and especially the close reading of long illustrative passages that student readers may feel tempted to skip over), creates an opening for metaconversations about the ethical importance of paying attention. I introduce students to Plumwood's concept of backgrounding, defined as a "perceptual politics of what is worth noticing, what can be acknowledged, 'foregrounded' and rewarded as achievement, and what is relegated to the background" (57). Plumwood's term helps students make connections between the various forms of social and environmental invisibility they may have encountered or engaged in and allows them to bring prior knowledge and lived experiences into their discussion of Kingsolver's novel.

In *Flight Behavior*, "seeing" is neither natural nor automatic. It is a dynamic, context-dependent process that must be learned. When the protagonist's toddler plays with a plastic rotary telephone, the little girl does not see a "phone" (those are flat and pocketable); instead, she turns the toy into a pet or an instrument. Dellarobia marvels at the perceptual gap between her daughter and herself: "She'd seen something so plainly in this toy that was fully invisible to her child, two realities existing side by side" (134). Similarly, views of the monarch butterflies differ dramatically across the novel's characters: Dellarobia's father-in-law sees pests; her mother-in-law sees familiar "King Billies" (118); the pastor sees a "beautiful vision of God's abundant garden" (72); and so on. Dellarobia's understanding of the monarchs, and of the more-than-human world in general, evolves rapidly as she learns scientific vocabulary and methods of observation. For example, she only notices that watercress is "frozen to blackness in the air above but still green underwater" (249) after she learns the technical term for this phenomenon from Ovid Byron: "She had heard him say the word *thermocline*, and now she could see that, too" (250). The pedagogical power of Kingsolver's text lies in the fact that the reader learns alongside the protagonist, acquiring not only new terminology but also metacognitive insight into the processes of perception and learning. Dellarobia's "I know the term, therefore I see the thing" epiphany comes in handy when students are introduced to new concepts like the Anthropocene and its polemical counterconcepts: Jason Moore's "Capitalocene," Haraway's "Chthulucene," and Yusoff's "billion Black Anthropocenes." Reflecting on how knowing the word changes the world becomes part of Kingsolver's expansive lesson and opens opportunities for student discussions and metacognitive epiphanies.

Ecosystemic Awareness

Across her narrative development, Dellarobia hones the ability to discern systemic connections. For example, she initially thinks of a butterfly as a singular, bounded organism: an insect with pretty wings. (This straightforward view was shared by most of my own students, as an in-class definition-writing exercise revealed.) Ovid shows Dellarobia that an animal cannot be considered in isolation from its environment. What makes a butterfly, he argues, is not "the physical body" but "the sum of its behaviors," including "its community dynamics," its "interactions with other monarchs, habitat, the migration, everything" (317). This expanded vision allows Dellarobia to see monarchs as a complex system, spanning continents and body forms. Similarly, she begins to see her own geographical location (Feathertown, Tennessee) within broader bioregional, continental, and planetary systems. The butterfly colony's anomalous landing in her backyard (instead of the mountains of Michoacán, Mexico) signals a systemic dysfunction. The same is true for the climate disasters Dellarobia had complacently assumed would happen only elsewhere, now arriving on her doorstep.[1]

As Dellarobia connects the dots between global warming and floods, deforestation and mudslides, and the Turnbows' secret stash of DDT and the neighbor boy's cancer (as well as her own traumatic miscarriage), she becomes aware of what Nixon calls "slow violence": "violence that occurs gradually and out of sight, a violence of delayed destruction that is dispersed across time and space, an attritional violence that is typically not viewed as violence at all" (2). Dellarobia also begins to locate herself within broader historical and socioeconomic contexts: the late-twentieth-century move to global capitalism that left behind skilled American workers like her parents; the debt pressures that force her family to put short-term profits over long-term sustainability; the relentless cycle of consumption that casts her as the buyer of shoddy products made by disenfranchised laborers on the other side of the planet. Shopping at the dollar store, Dellarobia realizes "there had to be armies of factory workers making this slapdash stuff, underpaid people cranking out things for underpaid people to buy and use up, living their lives mostly to cancel each other out. A worldwide conspiracy of bottom feeders" (159). The protagonist's educational journey brings her from an atomized understanding of material and social reality to a relational one. This expanded, systemic view of Dellarobia's identity creates an opportunity for students to explore their own identities as subjects within ecosocial systems.[2]

One quick and easy method of introducing students to the novel's expanding vision is an analysis of the first seven chapter titles: "The Measure of Man," "Family Territory," "Congregational Space," "Talk of a Town," "National Proportions," "Span of a Continent," and "Global Exchange." As students notice the outward-moving pattern, I invite them to reflect on how their own understanding of personal experiences expands by zooming out to encompass familial, communal, local, national, and global structures. In other reflective exercises, students might consider the material origins and transformations of the everyday objects before them (a cell phone, a sneaker, a cup of coffee) or contemplate the porousness of their own biological bodies. Kingsolver's reflection on how her writing was influenced by her study of science provides a useful starting point: "I'm well aware of my dependence on other kinds of beings. Every breath I inhale was manufactured by a leaf; all that I eat was once alive, bent on its own survival and reproduction" (Wilkinson 38). These exercises in scalar flexibility invite students to think beyond binaries (self/environment, human/ animal, nature/culture, local/planetary) and to begin developing more flexible, relational, ecosystemic conceptual frameworks.

Dialogic Communication

While Kingsolver's bildungsroman focuses primarily on the education of Dellarobia, the expert Ovid Byron has much to learn as well. By setting up a series of exchanges between Dellarobia (who learns about the science of climate change) and Ovid (who learns why people refuse to accept the science), Kingsolver invites readers to consider the role of affective, linguistic, and social schemas in the construction of knowledge and belief systems. The two characters' ongoing conversation demonstrates Kingsolver's commitment to representing multiple—sometimes conflicting—points of view within a single narrative frame.

The protagonist's many conversations with activists, scientists, journalists, and community stakeholders reveal a heteroglossia of positions that make it difficult for the reader to maintain a singular view on how to understand and respond to climate change. In a memorable scene, Dellarobia is accosted by an activist, Leighton Akins, bent on presenting her with his carbon-footprint-lowering "Sustainability Pledge" (327). As Akins goes down his list of action items, it becomes clear that they all assume a comfortable, middle-class lifestyle unattainable for Dellarobia (carry reusable containers for restaurant leftovers, buy energy-efficient appliances, fly less). Such universalizing assumptions represent what Ramachandra Guha and Joan Martinez-Alier call "'full-stomach' environmentalism" (xxi). The scene provides an entry point for readers on either side of the privilege divide, offering, depending on where one sits, a satirical mirror or an acknowledgment of injustice. Through such strategic dialogism, Kingsolver's narrative resists one-dimensional approaches to environmental questions, inviting readers to reconsider their own positionality through the critical lens of another.

Flight Behavior offers a wealth of dialogue scenes ready for in-class dramatic reading—a simple but effective classroom tool that makes space for complexity and tension without pushing for facile resolutions. More active classroom exercises include low-stakes creative writing prompts that ask students to imagine dialogic exchanges between characters in engaging formats (text message threads, phone conversations, etc.) that highlight the diversity of the novel's subject positions. Instructors looking for a more immersive application of the dialogic principle could structure parts of their course around elective role-playing scenarios in which students may choose to represent the positions of various stakeholders in contemporary climate change conversations: activist organizations and alliances (e.g., the Indigenous Environmental Network, 350.org, Sunrise Movement, and Fridays for Future), international bodies (like the Intergovernmental Panel on Climate Change or the European Union), corporate entities (like Tesla or BP), elected representatives, journalists, and so on. Another creative format is the climate change pop-up, which invites students to design an interactive, community-facing public event.[3] These dialogic activities encourage students to appreciate the multidimensionality of climate change, allowing them to imagine diverse modes of climate communication and action.

Above All Else: Do No Harm

The jury is still out on cli-fi's ability to represent—let alone remedy—the climate crisis. Critics question the genre's capacity for conveying the superhuman timelines and scales of climate change (Clark; Ghosh; Taylor). Champions point out its ability to localize global processes, traverse time and space, and devise detailed speculative scenarios that invite ethical reflection (Trexler; Mehnert; Bracke). Qualitative and quantitative research into cli-fi's capacity for influencing readerly attitudes is ongoing (Małecki et al.; Schneider-Mayerson, "Influence").

Whatever we hope our students gain from the study of climate fiction—be it scientific literacy, appreciation for a new literary mode, or conceptual overhaul—we must pay attention to the ways in which cli-fi can replicate injurious ideas and omissions. In this respect, Kingsolver's novel is notable not only for what it does but also for what it does not do. It does not peddle "disaster porn" or feed our addiction to spectacular violence. It does not perpetuate racist myths of neosavagery and precolonial "wilderness." It rejects masculinist myths of man-against-nature heroism. It resists the instrumentalization of the more-than-human world, including the commodification of "nature" as object of aesthetic pleasure. It questions—albeit not forcefully—what Nixon calls "superpower parochialism": the centering of the American experience while ignoring the capacity of the United States to "rupture the lives and ecosystems of non-Americans" (34). In short, Kingsolver attempts to account for the intersections of class, race, gender, and environmental justice in ways rarely found in US climate fiction.[4]

These attempts are not uniformly successful. That Dr. Byron, a Black American scientist, would be unaware of US educational inequities stretches the fabric of Kingsolver's impeccable realism. Ovid's neocolonial origins (the US Virgin Islands) are relevant primarily as the source of his "fascinating accent" (104). The novel's affirmation of upward mobility does not extend to the Delgado family of Mexican climate refugees or to other migrant Latino laborers whose lives are acknowledged only in passing. In short, *Flight Behavior*'s white-dominant universe addresses structural racism only obliquely. These gaps and asymmetries need to become subjects of conversation in the climate fiction classroom, given that, as

Yusoff argues, the current crisis cannot be understood apart from questions of race:

> It is racialized matter that delivers the Anthropocene as a geologic event into the world, through mining, plantations, railroads, labor, and energy. While Blackness is the energy and flesh of the Anthropocene, it is excluded from the wealth of its accumulation. Rather, Blackness must absorb the excess of that surplus as toxicity, pollution, and intensification of storms. (82)

Cli-fi that fails to grapple with these historical and present realities falls short of the genre's full pedagogical potential.

While a careful reading of Kingsolver's *Flight Behavior* fosters epistemological attentiveness, ecosystemic awareness, and dialogic communication, it also necessitates conversations about who is centered and what is omitted in mainstream climate change stories—and about what it would take for these stories to become more intersectional and just. How do we make space for the new climate narratives and visions? How do we make meaningful connections across social, species, and category boundaries? This is the vital work that lies before us and our students.

Notes

1. Critical discussions of *Flight Behavior* emphasize Kingsolver's use of ecosystemic connections to illustrate "the emancipatory potential of science" (Strauss 347), promote an "ethics of connectivity" (Mayer 489), and help reconcile local and planetary scales (Lloyd and Rapson).

2. The "Social Identity Wheel" is used at many universities to prompt students' exploration of their identities in a classroom setting.

3. For a discussion based on my own experience, see Feilla and Mączyńska.

4. For critique of climate fiction's insufficient engagement with the climate justice framework, see Schneider-Mayerson, "Whose Odds?"

Works Cited

Bracke, Astrid. *Climate Crisis and the Twenty-First-Century British Novel.* Bloomsbury, 2018.

Clark, Timothy. *Ecocriticism on the Edge: The Anthropocene as a Threshold Concept.* Bloomsbury, 2015.

Feilla, Cecilia, and Magdalena Mączyńska. "Pop! Taking Learning beyond the Classroom through Multimodal Pop-Up Events." *Transformations: The Journal of Inclusive Scholarship and Pedagogy*, no. 30, vol. 1, 2020, pp. 35–46.

Ghosh, Amitav. *The Great Derangement: Climate Change and the Unthinkable.* U of Chicago P, 2016.

Guha, Ramachandra, and Joan Martinez-Alier. *Varieties of Environmentalism: Essays North and South*. Earthscan, 1997.

Haraway, Donna J. *Staying with the Trouble: Making Kin in the Chthulucene*. Duke UP, 2016.

Kingsolver, Barbara. *Flight Behavior*. HarperCollins Publishers, 2012.

Lloyd, Christopher, and Jessica Rapson. "'Family Territory' to the 'Circumference of the Earth': Local and Planetary Memories of Climate Change in Barbara Kingsolver's *Flight Behaviour*." *Textual Practice*, no. 31, vol. 5, 2017, pp. 911–31.

Małecki, Wojciech, et al. "Literary Fiction Influences Attitudes toward Animal Welfare." *PLOS One*, 22 Dec. 2016, https://doi.org/10.1371/journal.pone.0168695.

Mayer, Sylvia. "From an Ethics of Proximity to an Ethics of Connectivity: Risk, Mobility, and Deterritorialization in Barbara Kingsolver's *Flight Behavior*." *Amerikastudien / American Studies*, vol. 61, no. 4, 2016, pp. 489–505.

Mehnert, Antonia. *Climate Change Fictions: Representations of Global Warming in American Literature*. Palgrave Macmillan, 2016.

Moore, Jason, editor. *Anthropocene or Capitalocene? Nature, History, and the Crisis of Capitalism*. PM Press, 2016.

Nixon, Rob. *Slow Violence and the Environmentalism of the Poor*. Harvard UP, 2011.

Plumwood, Val. "Decolonizing Relationships with Nature." *Decolonizing Nature: Strategies for Conservation in a Post-colonial Era*, edited by William M. Adams and Martin Mulligan, Earthscan, 2003, pp. 51–78.

Schneider-Mayerson, Matthew. "The Influence of Climate Fiction: An Empirical Survey of Readers." *Environmental Humanities*, vol. 10, no. 2, 2018, pp. 473–500.

———. "Whose Odds? The Absence of Climate Justice in American Climate Fiction Novels." *ISLE: Interdisciplinary Studies in Literature and Environment*, vol. 26, no. 4, autumn 2019, pp. 944–67.

"Social Identity Wheel." *Inclusive Teaching*, U of Michigan, 2021, sites.lsa.umich.edu/inclusive-teaching/social-identity-wheel/.

Strauss, Kendra. "These Overheating Worlds." *Annals of the Association of American Geographers*, vol. 105, no. 2, 2015, pp. 342–50.

Taylor, Jesse Oak. "The Novel after Nature, Nature after the Novel: Richard Jefferies's Anthropocene Romance." *Studies in the Novel*, vol. 50, no. 1, spring 2018, pp. 108–33.

Trexler, Adam. *Anthropocene Fictions: The Novel in a Time of Climate Change*. U of Virginia P, 2015.

Wilkinson, Crystal. "Barbara Kingsolver." *Appalachian Heritage*, vol. 42, no. 4, 2014, p. 38.

Yusoff, Kathryn. *A Billion Black Anthropocenes or None*. U of Minnesota P, 2018.

Teresa A. Goddu

Contemporary US Climate Fiction

For the past seven years I have been teaching a twenty-first-century US climate fiction course. It has been among the most rewarding teaching experiences of my career, and many of my students tell me that it has been similarly rewarding for them. Students connect deeply with the ethical imperatives that draw me to teaching contemporary climate fiction—namely, to foreground the urgency and inequities of the climate crisis. Their desire for truth-telling about the climate crisis and their need to understand its complexities and ramifications are palpable in our discussions.

As an upper-level English course that also counts for the climate studies major, the class addresses a range of students and, hence, has multiple goals. The first is to define the emerging genre. While literature has long been attentive to climate concerns, this course focuses on its contemporary manifestations as written under a new awareness of anthropogenic climate change. Over the last decade or so, climate fiction has become a focus of literary production and readerly interest. Many contemporary American authors—among them Jesmyn Ward, Lauren Groff, Barbara Kingsolver, Cormac McCarthy, N. K. Jemisin, Jeff VanderMeer, Colson Whitehead, Ben Lerner, Kim Stanley Robinson, Louise Erdrich, and Paolo Bacigalupi— have written climate fiction; *Goodreads* has several lists devoted to climate

150

fiction as well as an online book group; and *Twitter* has a lively feed devoted to the topic (#clifi). Coalescing into an identifiable literary form, climate fiction has become a significant feature of the contemporary literary landscape.

Understanding climate fiction as a broad and flexible category, the course, focused on novels and short stories, surveys both emerging and established authors as well as a range of literary modes, including realism, satire, and fantasy. It begins by asking the question "What is climate fiction?" and directs students to the wide array of answers included in the Amazon original story collection *Warmer*. This mini-anthology, explicitly advertised as climate-focused, provides students an entry point into the genre's range of themes, tones, and storytelling strategies. In discussing how, and even if, the stories fit under the designation "climate fiction," students are asked simultaneously to recognize the diversity of the genre—it no longer consists only of "disaster stories set in the future" (Ghosh 72)—and trace similarities among the texts. The goal of this exercise is to begin to articulate the genre's shared codes and conventions while also underscoring the difficulty of any easy definition. The exercise also allows for a conversation about how institutional contexts—in this case constructed by Amazon, as both a digital reading platform and publishing house—shape readers' expectations and engagement with texts. By beginning with the problem of genre, especially the difficulty of describing an emergent one, I introduce students to questions of how literary genres are formed.

Throughout the semester, I encourage students to take part in the process of genre-making by inviting them to contribute a weekly entry to the class's motif library. Housed on a platform that allows students to read one another's work (I have used my university's course management system as well as *Slack*), the motif library becomes an archive of the shared features and techniques that students trace across the texts we read. It can be set up in several ways: each student can be assigned a single motif to follow throughout the semester (e.g., the storm, the child, the shelter, the flood myth) and to synthesize all their entries into an argument about the motif's meaning and role in the genre in a short paper; or they can write on a different motif each week, adding new motifs to the library as they see them or building on previous entries. In both cases, the class uses the motif entries to map the contours of the field. The motif library keeps the question of genre foregrounded in our discussion of individual texts even as it weaves together our conversations across texts. Motif entries also build students' interpretive abilities and writing skills. These three-to-five-sentence nuggets of critical analysis train students to distill and deepen their understanding of a single aspect of a text and to draw connections among works. Modeled on Eric Hayot's "uneven U," where a paragraph works from conceptual claim to evidence and analysis and then back out to a broader conclusion, motif entries reinforce both close reading and analytical writing (59–73). This small-scale reading and writing practice is helpful to all students but is particularly useful to students from disciplines outside English who are new to these methods. The motif library also serves as a generative springboard for weekly class discussion as well as a way to conclude the course by synthesizing its various through lines.

Another way to foster students' participation in constructing the genre and to enhance their skills is to work with book reviews. In studying a text's reception by reading a range of reviews (I refer my students to the website *Book Marks* (bookmarks.reviews), which excerpts and collates reviews of contemporary literature), students see how a work is framed (or not) as climate fiction and are provided models for their own critical assessments. Short presentations on the reception history of a book can enhance class discussion by situating student responses in a larger context, and writing reviews can build critical voice. Book reviews also allow students to experience how readers as much as writers shape genres.

The second goal of the course is to situate the genre within a broader discourse about climate. I think of this shift as moving from a literary to an environmental humanities framing of the fiction. I do this by pairing a keyword with each text. The keyword provides a conceptual lens through which to examine the text and introduces students to important ideas in the broader interdisciplinary field of environmental and climate studies. For instance, I teach Karen Walker's young adult novel *The Age of Miracles*, which represents the climate catastrophe slantwise by showing what happens to the environment as well as society as the earth's rotation begins to slow, through Rob Nixon's concept of slow violence. We read the mysterious and ever-expanding Area X of Jeff VanderMeer's *Annihilation* as a manifestation of Timothy Morton's formulation of climate as a hyperobject, and we examine the southern wasteland of Cormac McCarthy's *The Road*, littered with the debris of fossil fuel and consumer capitalism and mired in America's history of plantation slavery, through the twinned terms *Capitalocene* (Moore) and *Plantationocene* (Haraway, "Anthropocene"). While I foreground a keyword with each text, the class reengages these concepts as they appear in other texts. For example, we discuss Jesmyn Ward's *Salvage the Bones* through the framework of environmental racism and environmental justice—studying the disproportionate effect of the climate crisis on poor as well as Black and brown communities—and note how the long lead-up to Hurricane Katrina emphasizes the slow violence of the Batiste family's everyday life over the more spectacular force of the eventual storm. Similarly, our conversation about how Ben Lerner's *10:04* highlights the role the commons plays in addressing the climate crisis also reflects on the varied ways the collective functions as both a salve and potential solution in climate fiction. Keywords, then, simultaneously bring a focus to our discussion of individual texts, build students' conceptual vocabulary and understanding, and situate the class's conversations in a wider, more interdisciplinary context.

The assignment that I connect to the keyword approach is the news article. Over the course of the semester, students are required to locate five news reports that relate to one of our terms or texts. In writing a short entry about each article, students apply their understanding of the keyword and expand their reading of the text to engage current issues. Students are eager to see and make connections to their own life and world, and news items provide them with a springboard for important conversations such as whether their generation should have children, how oil spills act like hyperobjects, or how rewilding, even just of one's yard, can benefit entire ecosystems. Thus news articles allow for a robust interdisciplinary and real-world framework to inform class conversation. In a more advanced course, students can also research and write short keyword entries much along the lines of the living lexicon located in the journal *Environmental Humanities*.

The third goal of the course is to explore literature's ability to confront the climate crisis. This conversation requires the class to reflect on the work literature does to transform society and to understand, as Amitav Ghosh writes, that "the climate crisis is also a crisis of culture" (9). Most broadly, we discuss how literature can play a role in creating social change. More specifically, we consider how climate fiction provides a place to comprehend, process, and reimagine the climate crisis. Through its ability to name and frame, climate literature makes visible and clarifies the complexities of the crisis. Through its metaphors and metonymies, it makes climate's unfathomable enormities more manageable and, hence, understandable: for instance, by depicting, as *The Age of Miracles* does, planetary change through the more familiar guise of adolescent transformation or by encapsulating, as *10:04* does, the global food chain in a single jar of instant coffee. Through its many modes of storytelling, it engages readers as it asks them to "stay with the trouble," to borrow Donna Haraway's term (*Staying*)—to notice, grasp, interrogate, and care about the crisis.

By registering the climate crisis as an affective problem, climate fiction also opens a space for readers to negotiate the difficult emotions of dread, despair, guilt, loneliness, and grief that accompany the existential crisis of extinction as well as build their capacity for hope, beauty, courage, and joy in the face of it. Lauren Groff's protagonist in "Boca Raton" lives in a sleepless state of climate dread that is itself a displacement for the grief and guilt she feels about her partner's likely death and her daughter's dark future, whereas Ben Lerner's narrator in *10:04* imbibes moments of sublime beauty despite his deep anxiety. In *Salvage the Bones*, the children's grief over the loss of their mother—an often-used metaphor for the destruction of Mother Nature in climate fiction—evolves into a deep sense of care and community for themselves and the nonhuman world as they keep watch over one another and await their lost dog's return at the end of the novel. Climate fiction acknowledges what we may not be ready to feel or face even as it builds empathy for ourselves, others, and the world around us.

We also consider how climate literature allows us to imagine otherwise: how it provokes us to rethink our assumptions and envision alternatives. Climate fiction can help readers access other ways of seeing, knowing, and being. The biologist's acceptance of Area X's radical unknowability in *Annihilation* as well as her transformation to a new state serve, for instance, as a model for the reader's own alteration in the act of reading. I encourage students, both through the texts I choose and through the issues I highlight, to see climate fiction as a space of radical possibility rather than apocalyptic determinism. Climate fiction, I underscore, provides not only catastrophic storms but also the means to weather them. It surveys the dark but also provides enough light to guide the way forward. It chronicles disaster and destruction even as it marks moments of minor miracles—a single egg "whole and mute, holding all the light of dawn in its skin" (100) that survives the hurricane in Groff's story "Eyewall," for instance. Moreover, climate fiction feeds our creative capabilities and provides the comfort of comradery. As Lerner's narrator looks at the totaled city at the end of the novel, he directly addresses the reader, stating, "I know it's hard to understand / I am with you, and I know how it is" (240).

We return to the question of literature's value in the moment of climate crisis regularly throughout the semester, since much of the fiction asks this question. Jenny Offill's *Weather*, for example, presents itself as a prepper manual that teaches readers a range of survival skills, from how to make oil lamps with cans of tuna to how to tell jokes. Louise Erdrich's *Future Home of the Living God*, written as Cedar's journal to her unborn child, argues that storytelling is a mode of survivance: it makes meaning and affirms life even as the world unravels. Other texts, such as Richard Powers's *The Overstory*, explicitly discuss art as a mode of environmental activism. Besides embracing art as a mode of survival, climate fiction also articulates its limitations. The rotting journals the biologist discovers in the lighthouse in *Annihilation*—a version of her own notebook and, hence, of the novel itself—communicate nothing and seem useful only as compost. Roberto, at the end of *10:04*, dismisses the book he and Ben have written in favor of making a movie. Climate fiction, then, constantly interrogates its own ability to combat the climate crisis.

The assignment that I pair with this learning goal is a creative final project. Students build a bug-out bag from items they have encountered in the literature we have read, imagining what they will need not only to survive but to thrive in a world defined by the climate crisis. Ending the course with a text like Lydia Millet's *A Children's Bible*, which features young people leaving their parents' house and paradigms behind as they forge a new society, can serve as a good segue into this assignment. Adapted from Farah Jasmine Griffin's COVID-19 assignment, which stretches students' imaginations and encourages them to create their own art, this project invites students to use the class texts as a resource and springboard for their own envisioning of a world transformed by climate change. The project opens with a five-hundred-word artist's statement in which students establish the issue, purpose, value, or action that guides the building of their bag. This encourages them to assemble a carefully curated collection of items rather than a grab bag. The requirement that they add one text from the semester to their bag as a map to guide them through the difficult terrain ahead further focuses their statement. They must then provide a list of ten items that will aid them in their journey, explaining the significance of each, as well as a list, also with short explanations, of five things they would leave behind. These items can range from the practical (the flare gun from *The Road* or the dried packets of ramen noodles from *Salvage the Bones*) to the abstract (the consumerism of *The Road*'s Coke bottle or the importance of memory as embodied in the photo of Esch's mother that her father saves from the storm in *Salvage the Bones*). The bag—which students are welcome to transform into other things, such as a vest (to keep them warm, cocoon them from anxiety, and provide pockets to store their items) or boat (a vessel in which they can weather the rising seas)—spurs students not only to imagine their own futures but to shape them. By seeing literature as a tool to be picked up and used, much like Rebecca Solnit's description of hope as an axe to be used to "break down doors with in an emergency" (5), rather than a lottery ticket to clutch, students become empowered to remake their worlds.

Throughout the course, I embrace pedagogical strategies that cultivate the capacities students need to face the climate crisis. Critical thinking, creativity, and collaboration are among the crucial skills my pedagogy fosters. In addition to providing a range of activities and assignments that develop these competencies, I use a portfolio system in which students can revise their work throughout the semester in response to peer review and teacher feedback. Structured around revision and reflection, portfolio writing allows students the space to experiment and take risks as well as the opportunity to increase their metacognition. I use group work of all kinds to build students' collaborative capabilities. The last week of class, for instance, is devoted to brainstorming the bug-out bags. Students quickly learn how much their work improves in dialogue with others. Finally, I work to create a culture of care and mutual aid in my classroom. For, as most climate fiction insists, the perils of the climate crisis are best addressed through the collective. By forging a connection and commitment to one another within our classroom, my students and I begin to build the communal infrastructures we will need to navigate the climate crisis.

Works Cited

Erdrich, Louise. *Future Home of the Living God*. HarperCollins Publishers, 2017.

Ghosh, Amitav. *The Great Derangement: Climate Change and the Unthinkable*. U of Chicago P, 2016.

Griffin, Farah Jasmine. "Teaching African American Literature during COVID-19." *Boston Review*, 25 May 2020, bostonreview.net/articles/farah-jasmine-griffin-teaching-african-american-literature-during-covid-19/.

Groff, Lauren. "Boca Raton." *Warmer*, Amazon, 2018.

———. "Eyewall." *Florida*, by Groff, Riverhead, 2018, pp. 84–100.

Haraway, Donna J. "Anthropocene, Capitalocene, Plantationocene, Chthulucene: Making Kin." *Environmental Humanities*, vol. 6, no. 1, May 2015, pp. 159–65.

———. *Staying with the Trouble: Making Kin in the Chthulucene*. Duke UP, 2016.

Hayot, Eric. *The Elements of Academic Style: Writing for the Humanities*. Columbia UP, 2014.

Lerner, Ben. *10:04*. Faber and Faber, 2014.

McCarthy, Cormac. *The Road*. Vintage, 2006.

Millet, Lydia. *A Children's Bible*. W. W. Norton, 2020.

Moore, Jason. *Capitalism in the Web of Life: Ecology and the Accumulation of Capital.* Verso, 2015.

Morton, Timothy. *Hyperobjects: Philosophy and Ecology after the End of the World.* Minnesota UP, 2013.

Nixon, Rob. *Slow Violence and the Environmentalism of the Poor.* Harvard UP, 2013.

Offill, Jenny. *Weather.* Alfred A. Knopf, 2020.

Powers, Richard. *The Overstory.* W. W. Norton, 2018.

Solnit, Rebecca. *Hope in the Dark: Untold Histories, Wild Possibilities.* Nation Books, 2004.

VanderMeer, Jeff. *Annihilation.* Farrar, Straus and Giroux, 2014.

Walker, Karen Thompson. *The Age of Miracles.* Random House, 2012.

Ward, Jesmyn. *Salvage the Bones.* Bloomsbury, 2011.

Robert P. Marzec

Margaret Atwood's *Oryx and Crake* and *The Year of the Flood* as Cli-Fi

Teachers of Margaret Atwood's *Oryx and Crake* and *The Year of the Flood* have at their disposal a wealth of information about climate change with which to enrich students' engagement with the novels. In the history of scientific knowledge production, no other event has been studied with such intensity and by so many researchers and institutions. The comprehensive assessments on climate change issued by the Intergovernmental Panel on Climate Change (IPCC), the world's leading authority on the topic, are compiled by experts from more than 80 countries, involving more than 830 lead authors, 1,000 contributors, and 2,000 reviewers. Sources like the World Wide Fund for Nature's *Living Planet Report* series, the annual reports of the International Union for the Conservation of Nature and Natural Resources, the publications of the Food and Agricultural Organization of the United Nations, the UN's Millennium Ecosystem Assessment report, and the United States Global Change Research Program's *Fourth National Climate Assessment* show the impact that climate change is having on habitats, biodiversity, and food security and how these changes are affecting cultures around the planet. Other, very different publications, produced by military and economic institutions and various political think tanks, reveal how climate change is often co-opted to serve specific ideological agendas. This is not even to mention the innumerable works that have appeared from scholars in the humanities and the natural and social sciences who have come to substantially theorize our Anthropocene age. This rich and heterogeneous set of sources can help students discover how Atwood's novels engage the considerable number of sociopolitical forces subtending climate change and other Anthropocene issues.

Oryx and Crake and *The Year of the Flood* productively complicate readers' understanding of climate change by situating the event within a series of historical, philosophical, political, and economic frameworks. The novels are set in the near future of North America (one guesses roughly midway through the second half of the twenty-first century), when climate change has come to impact not only those living in planetary hot spots but also those far from the front lines of ecological devastation. Despite their future setting, Atwood's novels are profoundly immersed in this contemporary nexus of forces that not only affects climate change but also influences the way our minds access this reality. As works of "speculative fiction"—a term Atwood has often applied to her work (*In Other Worlds* 6)—the novels creatively develop these forces, exploring their latent potentials to see how our future might unfold.

In the world of the future, as in today's world, environmental issues are governed by specific discursive formations. The key formations explored in *Oryx and Crake* are the discourses of military defense and security, neoliberalism and corporate development, bioengineering, and technological mastery. These influences are backgrounded in *The Year of the Flood*, which focuses instead on the three competing approaches to climate change and the environment: anthropocentric versus biocentric, religious versus scientific, and adaptative versus mitigative

In *Oryx and Crake*, Crake is a bioengineer who ends up running the biotech company RejoovenEsense. His friend Jimmy is something of an artist whose talents are put to use for marketing purposes. In *The Year of the Flood*, the two narrators, Toby and Ren, are adopted by the ecoreligious group God's Gardeners, run by Adam One and his brother Zeb. In each novel, life has come to be thoroughly militarized across the political and civil spectrum, echoing related momentums at work at the start of our own twenty-first century. The central military authority in the novels is the secretive CorpSeCorps, a security organization somewhat akin to the Pentagon and staffed with "men . . . on constant alert" (27). In *The Year of the Flood*, Adam One frequently reminds his followers that they are potentially under surveillance. The Gardeners receive an extensive environmental education on a daily basis, but they are taught orally and are warned that writing is "dangerous . . . your enemies could trace you through it, and hunt you down, and use your words to condemn you" (6).

The Gardeners pose a threat to the CorpSeCorps and to corporations like RejoovenEsense because their approach to the ecosystem is fundamentally biocentric, as opposed to the corporate world's highly technological, anthropocentric relation to the natural environment. This opposition reflects a principal and ongoing debate among environmentalists. In the field of environmental ethics, an anthropocentric approach is valued for pragmatic reasons and seen as a powerful stance in terms of shaping political policy, which is by default human-centered (see Norton; Light). In the ecological sciences and agroecology, this anthropocentric approach can be seen in characterizations of the importance of natural habitats and biodiversity in terms of the "ecosystem services" they provide for humans (Millennium, *Ecosystems* [World Resources Institute] 1). The manipulation of nature for human ends is manifest at its largest scale with the idea of "geoengineering": placing reflective particles in the upper atmosphere to cool the earth, fertilizing the oceans to increase plankton to create a carbon sink, and so forth (see Crutzen; Hamilton). Military institutions like the Defense Advanced Research Projects Agency (DARPA) and the classified JASON group also began exploring geoengineering during the twenty-first century's first decade (see Kintisch).

Opposed to these human-centered approaches, biocentrism or ecocentric holism extends moral status to all species and even to ecosystems, such as forests or freshwater systems (see Taylor; Agar). Many of the descriptions of the natural world offered by Adam One closely parallel, for instance, Aldo Leopold's ecocentric "land ethic": "The land ethic simply enlarges the boundaries of the community to include soils, waters, plants and animals, or collectively, the land" (204). During one of the many feast days celebrated by the Gardeners—"Mole Day . . . [the] Festival of Underground Life"—Adam One encourages people to consider "the Earthworms and Nematodes and Ants, and their endless tilling of the soil, without which it would harden into a cement-like mass, extinguishing all Life." Furthermore, Adam One consistently encourages his followers to deconstruct the culturally fabricated boundaries that have come to exist between human beings and other species. He invites people not to

> overlook the very small that dwell among us; yes, without them, we
> ourselves could not exist: for every one of us is a Garden of sub-visual
> life forms. Where would we be without the Flora that populate the in-

testinal tract, or the Bacteria that defend against invaders? We teem
with multitudes, my Friends—with the myriad forms of Life that creep
about beneath of our feet, and—I may add—under our toenails.
(Atwood, *Year* 160)

These and similar statements parallel efforts by deep ecologists such as
Arne Naess, Bill Devall, and George Sessions; the latter two make the
case for a "biocentric equality" that endows nature with "intrinsic worth"
(Devall and Sessions 76, 69).

The focus on genetic engineering in *Oryx and Crake* is another part
of the complex constellation of knowledge discourses making up our access to the environment. Many of the concerns addressed by Atwood in
the novel are taken up directly by the environmentalist Bill McKibben in
his 2003 book *Enough: Staying Human in an Engineered World* (Atwood
reviewed the book for *The New York Review of Books* the same year; see
"Arguing"). McKibben's book interrogates the ethical and cultural implications of germline genetic engineering. This relatively new form of genetic engineering targets the basic cells from which humans "germinate."
Germline modifications in genes affect the entire body and are handed
down through subsequent generations. Crake's genetic experiments in the
novel are of this kind. His attempts to generate a more environmentally
responsible human reflect ideas that are already making up the conversational background surrounding genetic modification: engineering embryos
to produce humans with supposedly higher IQs, happier dispositions, or
cutthroat executive tendencies (see McKibben). Through "interspecies
gene . . . splicing," Crake genetically engineers a new, environmentally
friendly species called the Crakers. Gentle, herbivorous humanoids of "all
available skin colors" (Atwood, *Oryx* 302), the Crakers are the new "floor
models" (Crake deceptively tells Jimmy) created for couples wishing to pur-
chase genetically designed, biocentric babies with "enhanced immune
system functions" (302, 303). Crake adds that they are also ideal for governments interested in designing a docile citizenry from the ground up
(304). Thoroughly in balance with the ecosystem, the Crakers, we eventually discover, have been designed by Crake to replace the human species, which he believes is inherently violent and greedy, predisposed to the
overconsumption of resources and the destruction of the environment.

This thematization of overconsumption situates the novels at a point
in time when the self-destructive effects of neoliberal development, with
its essential dependency on petrochemicals and its overconsumption of
planetary resources, could no longer be denied. Meanwhile, in the world

where the novels were being written, such evidence continued to grow. Predatory lending, speculative investment, and the collapse of the housing market led to the financial crisis of 2007–08. Unequal distribution of wealth and resources reached an all-time high: CEO salaries reached five hundred times the salary of the average worker in the United States, and less than one percent of the world's population was found to possess more wealth than the poorest sixty percent combined (see Harvey; Coffey). These disparities are reflected in the world of the novels, in which the wealthiest live in heavily securitized "Compounds"—walled estates owned and run by competing corporations—while the majority live in urban slums called "pleeblands" (Atwood, *Oryx* 27). The "top people . . . execs and . . . scientists" live in the Compounds, manufacturing weaponized "hardware, . . . software, [and] . . . hostile bioforms"—manipulations of nature that reflect the social system's general mechanistic and aggressive approach to the environment (26–27, 28). David Harvey's *A Brief History of Neoliberalism* productively details neoliberalism's aggressive origins and its impact on environments, which began with covert military actions orchestrated by US corporations and the US government against countries in Latin America in the 1970s. This link between corporate neocolonial development and the aggressive securitization of territories is reflected in the novels. The enigmatic Corp-SeCorps runs most of the polity, and its sovereign network extends across the entire social spectrum. The CorpSeCorps's internecine maneuvers parallel developments in the "security society" and the "state of exception" as articulated by Michel Foucault and Giorgio Agamben, respectively.

The effects of neoliberalism's mandate to destroy the welfare society and governmental environmental oversight appear in stark relief in the novels. Public services like health care, education, environmental regulation, and relief for the poor are nonexistent. The arts as they are commonly understood today have disappeared. What remains are institutions like the rundown Martha Graham Academy, at which the novel's narrator Jimmy hones his playful language skills, only to end up using his talents to sell products for AnooYoo. The humanities seem to have disappeared along with all critical efforts to address inequalities operative in the culture at large. The one specific literary text mentioned in *Oryx and Crake*—the *Norton Anthology of Modern Poetry*—is said to be owned by a "word person" working for RejoovenEsense: a "speechwriter, an ideological plumber, a spin doctor" (233).

The civil war character of this corporate-compound life reflects not only real-world neoliberalism but also a related force at work on the institutional structures subtending climate change: the maneuver by US military authorities to control the climate change narrative. Crake's effort to technologically master nature and to geoengineer a brighter environmental future constitutes a speculation that arises out of analogous efforts orchestrated by US military and security authorities. Beginning in the 1990s and continuing into the present, the various branches of the US Department of Defense began to take a serious interest in climate change and other environmental problems (see Marzec). The resulting initiatives helped bring the seriousness of climate change to a broader public but also attempted to steer civil society toward a militarized understanding of what climate scientists refer to as "adaptation" (preparing for inevitable ecological changes to come—that is, violent conflicts erupting over limited resources) and away from "mitigation" (efforts to reduce or prevent ecological degradation). Fully accepting the science of climate change, the Department of Defense pushed the Obama and Trump administrations to take serious action on the issue. However, the goal in mind with adaptation is not only to prepare for changes but also to commandeer the complexities and severity of climate change for the purpose of directing public action toward a fateful future understood to constitute a threat to national security.

The US military has a long history of being interested in environmental issues, which extends back to the Cold War era of the 1960s (see Hamblin). But linking climate change to national security (which also means corporate security) has its origins in works like Robert Kaplan's politically influential 1994 essay "The Coming Anarchy: How Scarcity, Crime, Overpopulation, Tribalism, and Disease are Rapidly Destroying the Social Fabric of Our Planet" and in the IPCC's third and fourth assessment reports (Intergovernmental Panel, *Climate Change 2001* and *Climate Change 2007*). These reports revealed a strong scientific consensus regarding the seriousness of climate change and stated in no uncertain terms that climate change was the result of the human use of fossil fuels. Even though the US government did little to address climate change during the Clinton administration and continued to officially deny climate change during the Bush administration, the Department of Defense and security institutions like the Center for a New American Security became increasingly interested and vocal about climate change, characterizing it as a "threat multiplier" more dangerous than the war on terror (*National Security* 6). The police state of the novels not only reflects these developments but also bears an uncanny resemblance to one of the "future scenarios" developed two years later by the United Nations Millennium Ecosystem Assessment Team. Called "Order from Strength," the scenario presents a "regionalized and fragmented world, concerned with security and protection, emphasizing primarily regional markets, paying little attention to public goods, and taking a reactive approach to ecosystem problems." Economic growth rates are "at their lowest, particularly . . . in developing countries," and the gap between rich and poor countries, and between rich and poor people within countries, is at its most extreme in recent history, with the wealthy living in walled enclaves and the poor increasingly pushed beyond the point of survival (Millennium, *Ecosystems* [Island Press] 4).

I do not have the space here to enumerate the many other environmental developments that Atwood explores. I will note instead that climate change is never mentioned by anyone in either of the novels, not even the narrators, despite the fact that people are living through increasing heat stress, dealing with drought, facing food insecurity, and confronting a host of other environmental obstacles. The lack of a specific reference to climate change as its effects permeate the world of the novels situates readers in a unique and instructive position: it gives them a sense of how such a perverse reality could come to be so easily accepted by a population as normal. This narrative strategy, which I shall call here speculative normalization, recasts the worst environmental conditions as part of an everyday existence. This serves a double purpose. First, it allows us to reflect on our current historical situation, a new normal that is grim for many. Second, it allows us to immerse ourselves in an even more ominous future that we are already coming to inhabit. When I first started teaching these novels, climate change was not yet a topic often discussed in the popular press. I found myself spending considerable time presenting students with the facts of what had recently become clear to many researchers. This has fundamentally changed. In the last dozen years, students have read, heard, and even directly seen the effects of climate change unfolding. Still not adequately discussed, however, and very often unacknowledged, are the many ways in which our potential responses to this event are being absorbed by powerful pressures at work in our culture. I have attempted here to detail some of these pressures and to demonstrate that Atwood's novels show how the forces surrounding the event of climate change may come to overwhelmingly define how we think about, access, and live in the troubled environments of the Anthropocene.

I encourage students to interrogate these relationships through an activity that centers on Adam One's call to deconstruct the boundaries that have come to exist between humans and nonhumans—specifically, his reimagining of the human as a "Garden of sub-visual life forms." He deepens this nonanthropocentric contextualizing of human subjectivity earlier in the novel when he challenges the Protagorean dictum that "Man [is] the measure of all things" (Atwood, *Year* 40), thus introducing an awareness of multispecies measurement and meaning making. I expand upon this by discussing Jakob von Uexküll's concept of the *Umwelt*, or species-specific environment: the awareness that each species within an ecosystem has its own world of elements that it finds significant and with which it interacts. Each species is deeply embedded and living in a set of moving interrelations with other species, but each has its own world, its own "environment," which is intrinsically alien to others. Most of all, the environments of nonhumans have become alien to humans.

Prior to introducing the concept of the *Umwelt*, I set up a series of simulated *Umwelten* in the classroom before students arrive. The *Umwelten* are each based on the various significances and habitats of six different species: beavers (*Castor*), hummingbirds (*Trochilidae*), oak trees (*Quercus*), rusty-patched bumble bees (*Bombus affinis*), sea otters (*Enhydra lutris*), and humans (incorporating visual, aural, and movement cues meaningful to each species). Students enter a dimmed classroom to find gobo lights of various colors illuminating specific spaces in the room, coupled with a variety of sounds specific to the different species' environments (running water, ocean waves, various insect sounds, and traffic noises). They must explore the room to "find" the five nonhuman and one human worlds. The "species" themselves are actually hidden slips of paper conveying specific information (about a species' particular habitat, what it needs in order to survive and flourish, its relation and importance to other species and the overall ecosystem, a brief history of the human colonization of that species, and what would happen if that species were to become extinct). To simulate the dangers of anthropocentric measurement and understanding, and the impact that humans can have on the material worlds of nonhumans, each "species" is surrounded by thin string threads that are difficult to see. Accessing a species means that students find themselves unexpectantly caught in these threads. Moving too quickly can result in the destruction of a species' habitat. In order to not damage a habitat, students must attend to unaccustomed visual and aural cues. The goal here is to simulate an encounter with a species on its own terms—in other words, immersing oneself in a species' peculiar *Umwelt*. Once all the species are discovered, students share information about each. The exercise ends with a discussion focusing on how Atwood explores the potentialities of generating less destructive human and nonhuman relations (and nonrelations) and, more generally, how humankind might expand its limited *Umwelt* to encounter different concepts of value and different forms of intelligibility.

Works Cited

Agamben, Giorgio. *State of Exception*. Translated by Kevin Attell, U of Chicago P, 2005.

Agar, Nicholas. *Life's Intrinsic Value: Science, Ethics, and Nature*. Columbia UP, 2001.

Atwood, Margaret. "Arguing against Ice Cream." *The New York Review of Books*, 12 June 2003, nybooks.com/articles/2003/06/12/arguing-against -ice-cream/.

———. *In Other Worlds: SF and the Human Imagination*. Doubleday, 2011.

———. *Oryx and Crake*. Random House, 2004.

———. *The Year of the Flood*. Random House, 2010.

Coffey, Clare, et al. *Time to Care: Unpaid and Underpaid Care Work and the Global Inequality Crisis*. Oxfam, 2020.

Crutzen, Paul. "Albedo Enhancement by Stratospheric Injections: A Contribution to Resolve a Policy Dilemma?" *Climatic Change*, vol. 77, 2006, pp. 211–19.

Devall, Bill, and George Sessions. *Deep Ecology: Living as if Nature Mattered*. Gibbs Smith, 1985.

Foucault, Michel. *Security, Territory, Population: Lectures at the Collège de France 1977–1978*. Edited by Michael Senellart, translated by Graham Burchell, Palgrave, 1978.

Hamblin, Jacob Darwin. *Arming Mother Nature: The Birth of Catastrophic Environmentalism*. Oxford UP, 2013.

Hamilton, Clive. *Earth Masters: The Dawn of the Age of Climate Engineering*. Yale UP, 2013.

Harvey, David. *A Brief History of Neoliberalism*. Oxford UP, 2005.

Intergovernmental Panel on Climate Change. *Climate Change 2001: Synthesis Report: Contribution of Working Groups I, II, and III to the Third Assessment Report of the Intergovernmental Panel on Climate Change*. Edited by R. T. Watson and the Core Writing Team, Cambridge UP, 2001.

———. *Climate Change 2007: Synthesis Report: Contribution of Working Groups I, II, and III to the Fourth Assessment Report of the Intergovernmental Panel on Climate Change*. Edited by the Core Writing Team et al., IPCC, 2007.

Kaplan, Robert. "The Coming Anarchy: How Scarcity, Crime, Overpopulation, Tribalism, and Disease Are Rapidly Destroying the Social Fabric of Our Planet." *The Atlantic*, Feb. 1994.

Kintisch, Eli. "DARPA to Explore Geoengineering." *Science*, 14 Mar. 2009, www.sciencemag.org/news/2009/03/darpa-explore-geoengineering.

Leopold, Aldo. *A Sand County Almanac*. Ballantine, 1986.

Light, Andrew. "Contemporary Environmental Ethics: From Metaethics to Public Policy." *Metaphilosophy*, vol. 33, no. 4, 2002, pp. 426–49.

Marzec, Robert P. *Militarizing the Environment: Climate Change and the Security State.* U of Minnesota P, 2015.

McKibben, Bill. *Enough: Staying Human in an Engineered World.* Henry Holt, 2003.

Millennium Ecosystem Assessment. *Ecosystems and Human Well-Being: Biodiversity Synthesis.* World Resources Institute, 2005, www.millennium assessment.org/documents/document.354.aspx.pdf.

———. *Ecosystems and Human Well-Being: Scenarios.* Edited by Steve Carpenter et al., Island Press, 2005, www.millenniumassessment.org/en/Scenarios .html.

Naess, Arne. "The Shallow and the Deep, Long-Range Ecology Movement: A Summary." *Inquiry*, vol. 16, nos. 1–4, 1973, pp. 95–100.

National Security and the Threat of Climate Change. Center for Naval Analysis Corporation, 2007, www.cna.org/reports/2007/national-security-and-the -threat-of-climate-change.

Norton, Bryan. *Towards Unity among Environmentalists.* Oxford UP, 1993.

Taylor, Paul W. *Respect for Nature: A Theory of Environmental Ethics.* Princeton UP, 1986.

Jason de Lara Molesky

Cli-NoFi: Reading and Writing Creative Climate Nonfiction in a Prison Classroom

Writers of creative climate nonfiction—what we might call *cli-nofi*—have produced a wealth of transformative texts in the last several decades. These science-infused works can enrich many kinds of university syllabi across literary studies and the environmental humanities. This essay draws on my recent experience teaching climate nonfiction in a classroom of incarcerated college students. Principally, I outline pedagogical strategies for discussing cli-nofi and for asking students to write it as well. This type of authorial investment can animate abstract climate projections more effectively than textual analysis alone while also helping students integrate the complex, troubling affects that necessarily arise when we look with clear sight at the climate crisis. In what follows, I refer to several modes of cli-nofi but focus on what Rob Nixon calls "speculative nonfiction"—that is, works that use our best climate models to build textured histories of near-future worlds. Course modules that ask students to engage with this mode of literature as readers and as writers effectively join the narrative arts to the environmental sciences, seeking to marshal the capacities of both disciplines toward a more just and sustainable world.

The students I worked with in the late 2010s at one of the largest federal prisons in the United States harbored complex emotions about cli-

168

mate change. In keeping with the theme of our essay course, Climate, Race, and Democracy, I began the semester by asking what the term *climate change* meant to them. Hands crept up, and soon the conversation became general: polar bears, capitalism, wildfires, electric monster trucks. A few were convinced that climate change was a "hoax" or "fake science." But even they seemed moved as one student recalled the unease he had felt when the facility had lost power during Hurricane Sandy. "Society wouldn't care if we all ended up stranded here," he said. Heads nodded. Another talked of not knowing for weeks whether family members had survived. The students' scowls and anxious jokes made clear that they had emerged from the hurricane, as one man put it, "shook." For them, Sandy was not past at all but remained as a herald of the next, worse storm.

I simply let the students share their experiences, reticent to use their trauma to make imprecise analogies to other crises. Later, while driving out of the facility, passing the rooftop solar panels at the adjacent army base, I wondered whether my silence had been the right choice. Points of contact between personal chaos and global catastrophe, however tenuous, can deepen one's understanding of climate change's diffuse impacts on the marginalized. But I believed such insights would carry more weight if they sprang from the students themselves rather than from me, a volunteer instructor.

In my creative writing courses, I have had success employing works of creative climate nonfiction toward similar ends. This emerging genre, like other environmental nonfiction, takes many forms, from reportage, memoir, and travel writing to the extemporaneous oral testimonies of activists and frontline residents. Figures like Mohamed Nasheed, who as president of the Maldives defended his island state against the specter of inundation, have as much claim to representation in this canon as decorated literary writers like Lauret Savoy or Barry Lopez.[1] From a plethora of compelling cli-nofi materials, the prison-teaching team and I had chosen to center climate discussions around Elizabeth Kolbert's book-length work of reportage, *Field Notes from a Catastrophe.*

Published in 2006, *Field Notes* unequivocally demonstrates the reality of global climate change and shines light on the forces most responsible for political inaction. My students were engaged by the book's on-the-ground investigations and impressed by its range of arguments and evidence. They analyzed the structures of its claims and tried to apply them in their own critical essays. The students were objectively interested, in other words, but only rarely moved. Before long, I realized that other, supplementary cli-nofi texts might help them more deeply inhabit the Anthropocene world in which they daily found themselves confined.

I suspect that *Field Notes* and similar texts of its era have succeeded so terrifically in shaping public discourse that they no longer surprise. Many now entering college, including my younger students at the prison, have known about the principles of climate change since they were schoolchildren. Put simply, our sensoria have adapted, and our pedagogical strategies should adapt as well, moving from the abstract to the visceral and from data to narrative. Not long ago, we strove to convince the public that climate change is real. Now, thanks largely to the suasive force of voices like Kolbert's, most students not only accept climate projections; their nights are haunted by them. Our principal difficulty now lies in helping them modulate their affective responses while broadening their imaginative horizons in productive ways. *Field Notes* and similar texts should retain a place on many syllabi. Such elegant warnings, however, are often most effective when supplemented by texts that vividly, plausibly enact climate futures.

Speculative climate nonfiction performs this work. Rob Nixon, in his foundational essay on the subject, calls works of speculative cli-nofi "scientifically informed histories of the future." He writes, "Such anticipatory histories seek to counter a disastrous temporal parochialism unequal to the demands of the warmer, more insecure world. . . . In a spirit of anticipatory memory, writers, artists, and activists encourage us to own the future by inhabiting it in sample form." I concur with Nixon in stressing the essentially factual, if always contingent, character of speculative cli-nofi as a genre. Texts operating under this rubric keep within the bounds of current scientific knowledge and look out only a few decades, usually to around the year 2050. Given a certain emissions scenario, the range of outcomes over so short a time frame becomes narrow enough that speculative cli-nofi texts can be seen as offering conditionally truthful—that is, nonfictional— accounts of tomorrow.

The species of situated prolepsis that Nixon highlights takes inspiration from a rich science fiction tradition dating to at least the seventeenth century. But speculative cli-nofi is obviously different in that it thrives under the constraints of actuality. As Nixon writes, the genre's "imaginative channels" are "necessarily narrower" than cli-fi's, for cli-nofi "tries to sketch not just a possible future but a plausible one . . . constrained by science." On the level of detail, too, speculative cli-nofi revels in the factually grounded. For example, in one of the two scenarios depicted in Christiana Figueres and Tom Rivett-Carnac's *The Future We Choose*, clean water has grown so scarce that "the taps in nearly all public facilities are locked, and those in restrooms are coin-operated" (15). This line is followed shortly by a citational footnote. Such granular details may (still) seem fantastic, but they trade in climate science, not allegory or metaphor. By fostering the frisson of the mimetic in this way, speculative cli-nofi can drive student responses inaccessible to but often supportive of the complementary work of climate fiction.

Incarcerated people, against their will and without their knowledge, have often been conscripted as experimental subjects. I made sure to involve the students in my plans to supplement *Field Notes* with a short course module on reading and writing other climate nonfiction. The module was adapted from my creative writing courses, so most of the onus fell on them; the instruction time came out of the blocks we had previously devoted to in-class writing on their assigned essays. Several students had previously approached me about using these blocks to instead discuss additional texts. Still, I was surprised by how strongly they embraced the idea. I updated my materials and began the module a few weeks later.

Choosing among worthy texts is a difficult pleasure.[2] If *The Future We Choose* had existed at the time, I would have used it with my students— particularly its first three chapters, which justly receive pride of place in Nixon's analysis. I consider this opening section, which presents the two alternate histories of 2050 with minimal commentary, as probably the best cli-nofi text for most classrooms. It is short, detailed, and impactful. The high-emissions, business-as-usual scenario is chilling in its linkage of climate collapse and market fundamentalism, especially in regard to water and border security. And while one could argue that the solarpunkish, low-emissions scenario portrays its favored energy solutions a bit too uncritically, it also offers a laudable vision and serves as a generative foil.

Choosing from the works available, I considered using arguably the most decorated speculative cli-nofi text, Naomi Oreskes and Erik Conway's *The Collapse of Western Civilization: A View from the Future.* In a more advanced course, or toward the end of one devoted exclusively to environmental issues, it could make a very good choice. But whereas *Future* aims at a general readership, *Collapse* ventures into specialized concepts such as geoengineering and methyl hydrates without much explication, assuming a level of prior knowledge beyond what most undergraduates will likely bring to a course.

Instead, I turned to two speculative cli-nofi texts that were already part of my module and one adjacent work with which I wanted to experiment.

I began with Jeff Goodell's essay "Goodbye, Miami" (later adapted into the book *The Water Will Come*). Students typically respond powerfully to the opening, which provides a useful foothold for discussion:

> When the water receded after Hurricane Milo of 2030, there was a foot of sand covering the famous bow-tie floor in the lobby of the Fontainebleau hotel in Miami Beach. A dead manatee floated in the pool where Elvis had once swum. . . . The president, of course, said Miami would be back, that the hurricane did not kill the city, and that Americans did not give up. But it was clear to those not fooling themselves that this storm was the beginning of the end.

Asked for their impressions, students often note this section's matter-of-fact tone, precise imagery, and striking date markers. "Is this more like journalism or more like fiction?" I ask. This line of talk offers a good chance to introduce the concept of speculative cli-nofi and its relation to climate science. (Nixon's essay could make an excellent supplementary text in this vein.) Another useful exercise for "Goodbye, Miami" involves asking students to compare the opening paragraphs with the last two sections, in which Goodell discusses the same near-future Miami but in a more conventional manner, reverting to the future conditional tense in lines like "But more likely, the ocean will seep slowly into the city" or "The financial crisis could play out like this." At this point I like to ask, "Which section is more engaging to read?" And then, "Is one any more truthful than the other?"

The primary writing assignment of the module uses Goodell's opening section as a template and invites students to compose their own vision of a near-future place in relation to climate change. Confronting complex feelings by manifesting environmental loss on the page may prove constructive, even cathartic, for some students.[3] But rather than compelling everyone to inhabit this framework, I offer students the choice between a "goodbye" scenario, after the fashion of Goodell, or a "hello" scenario, one in which things turn out relatively well, whether through mitigation, adaptation, or controlled retreat. The essays can run from perhaps five hundred words to several thousand, depending on course goals. My prompt reads as follows:

> Writing as a literary journalist in the year 2050, tell about a specific place in terms of the physical and social impacts of climate change. Narrate the ways that human communities have changed, making sure to touch on salient events. Gesture toward broader vistas but situate us in the immediate scene. Use vivid details to disorient the reader and

defamiliarize the place. No magic technologies. The place must be on planet Earth. You may choose from two scenarios:

"Goodbye, Place X"—Global emissions continued to increase largely unabated until 2050: the "business-as-usual" scenario.

"Hello, Place Y"—Global emissions were reduced to the IPCC's 1.5°C target or even lower by 2050: the "successful mitigation" scenario.

The key elements I look for in these essays are plausibility, breadth of vision, and richness of detail. Regarding assessment, I find it unproductive to grade this assignment based on storytelling prowess or stylistic facility, especially in courses not focused on creative writing as such. While the essay may be accompanied in some classrooms by a brief artist's statement—a document in which the writer justifies their artistic decisions, in this case by referring to climate projections—I typically do not assign one. The idea is simply for students to imagine their way into the exercise as fully as they can, in the belief that writing—even more than reading alone—can help us, as Nixon puts it, "feel our way forward into the emergent worlds that our current actions are precipitating."

Inviting students to share their essays with one another, and devoting class time to collectively discussing them, can further promote this kind of inhabitation. I think of the process as a writing workshop without the workshopping—that is, without the critiques or the pressure to generate and remedy faults. Here, with this module, I want merely to afford students the chance to gain from one another's climatic visions.

The prompt may also be used alongside other speculative cli-nofi texts. In our prison classroom, for example, I asked the students to also read three of the short "Forty-Year Forecast" sections of Heidi Cullen's *The Weather of the Future*—on California's Central Valley, New York City, and the Arctic (115–48, 149–96, 227–60). Each assumes no significant emissions reductions and looks out to the year 2050 in increments; the New York section is typical in featuring imagined forecasts from 2013, 2014, 2017, 2027, 2039, and 2050. The general picture leans toward the dystopic, but the book also spotlights resilience measures, imagining, for instance, that storm surge barriers were gradually installed around the city following a Sandy-like hurricane in 2013. The students who had experienced the real storm seemed particularly taken with the book's prescient sketches of damage to subways and Rockaway beaches, which "nearly vanished" in Cullen's account (252). Cullen's vivid writing, some said, helped them sharpen their own.

Weather segued into the newest addition to the module, Elizabeth Rush's *Rising: Dispatches from the New American Shore*, winner of a 2019 Association for the Study of Literature and Environment Book Award. The text often reads like speculative cli-nofi, but the events it relates have actually happened. As Rush observes in the Staten Island chapter on coastal retreat after Hurricane Sandy, "[T]he future is, in many cases, already here" (120). Between the reportage sections that compose most of the text, Rush interpolates first-person oral testimonies of everyday people who have faced climate impacts. These acts of witness seethe through and inundate Rush's sober exposition, suggesting the work of Studs Terkel or Svetlana Alexievich. Nicole Montalto's words about her father's drowning at his waterfront, working-class home in Oakwood Beach, Staten Island, seemed to generate in many of the students the kinds of forceful affects that can surge beyond the mind's barricades and sea walls, carrying ideas like shells upon the crest. Her voice reads, at one point: "I don't like to talk about this stuff, I rarely ever do. *[Crying.]* Sorry. This is why I don't do interviews. But you're writing a book, and that's different because it will immortalize him" (Rush 104). In invoking the aura of writing at such a fraught moment, Montalto speaks to the elegiac capacity of literature, its increasingly important ability to memorialize the people and places lost to upheaval. For her and the other climatic harbingers in *Rising*, the speculative has become the actual.

In this spirit, I closed the module by asking students to write a letter about climate change to someone beloved who they expect will be living in 2050. The idea came from the Dear Tomorrow project, whose crowd-sourced website asks visitors to do the same but also requests that they pledge themselves in their letter to "bold climate action" on behalf of said beloved and then report on that action in a public post. I felt that many in the class might perceive such a pledge as a display of facile rectitude—an appeal for parole, empty of signification, meant to preclude further punishment. I asked of the students just the letter; its content I left to them.

I collected the letters for a completion grade but did not read them. Precise assessment of my students' intimate thoughts seemed far from the point. I brought the pages in my satchel to a quiet cabinet drawer. The next week I carried them inside: past the razor wire, the metal detector, the baggage scanner, and the double-keyed latch. I made my rounds and handed them back, leaving the words to their long, steady work.

Notes

1. For more on Nasheed, including his international climate advocacy and his famous underwater cabinet meeting, see *The Island President*.

2. Nixon's essay discusses many exceptional works that employ speculative cli-nofi strategies.

3. Exposure therapy often uses writing in this way to help patients overcome anxieties and phobias, and the same logic may apply to those struggling with the complex affects that result from climatic or environmental loss.

Works Cited

Cullen, Heidi. *The Weather of the Future: Heat Waves, Extreme Storms, and Other Scenes from a Climate-Changed Planet.* HarperCollins Publishers, 2010.

Figueres, Christiana, and Tom Rivett-Carnac. *The Future We Choose: Surviving the Climate Crisis.* Alfred A. Knopf, 2020.

Goodell, Jeff. "Goodbye, Miami." *Rolling Stone*, 30 Aug. 2013, rollingstone .com/feature/miami-how-rising-sea-levels-endanger-south-florida-200956/.

The Island President. Directed by Jon Shenk, AfterImage Public Media, 2013.

Kolbert, Elizabeth. *Field Notes from a Catastrophe.* Bloomsbury, 2015.

Nixon, Rob. "All Tomorrow's Warnings." *Public Books*, 13 Aug. 2020, public books.org/all-tomorrows-warnings.

Oreskes, Naomi, and Erik Conway. *The Collapse of Western Civilization: A View from the Future.* Columbia UP, 2014.

Rush, Elizabeth. *Rising: Dispatches from the New American Shore.* Milkweed, 2018.

Aaron Rosenberg

Genres of Deep Time:
Virginia Woolf's *Orlando*
and the Orbis Hypothesis

"When should we begin?" This is a question that perplexes the emerging genre of climate fiction and one that offers a productive starting point for teaching it. In my masters-level seminar on literature and ecology, my students and I explore how the problem of origination, a contentious issue in debates about the Anthropocene, speaks to broader epistemological challenges facing the environmental humanities. We consider how decisions about pinpointing the emergence of human agency on a planetary scale involve making significant choices about which knowledge systems we privilege and which forms of information should be considered relevant. Applying geological criteria carries certain risks, since stratigraphic boundary markers, or so-called golden spikes, are not, as Kathryn Yusoff reminds us, "immune to the narrative overtures" that "draw the world of the present into being and give race and shape to its world-making subjects" (24). Adhering to scientific paradigms, moreover, might increase demands on scholars in the humanities to justify the value of their research according to unsuitable or unworkable standards. What, after all, can literature say about climate change that would either supplement or subvert scientific findings? How can literary texts be read in relation to media like ice core

176

samples, atmospheric carbon measurements, and average temperature readings—to empirical data points that are being modeled using ever more powerful computational methods?

Rather than attempting to resolve these questions, my course aims to linger with them by examining how representational strategies developed in fiction inform the narrative structures resident in the concepts of climate change and the Anthropocene. By encouraging students to read these concepts as texts, I am asking them to engage with the idea that, as Dana Luciano puts it, the "Anthropocene offers climate change not just periodicity but narrativity. And like any well-told story, it relies on conscious plotting and the manipulation of feeling." Because my students are already well equipped with skills in textual analysis, I find that pursuing this approach allows us as a class to draw incisive connections between seemingly disparate sources and to trace temporal shifts that exceed conventional boundaries of discipline and period. For my purposes as an instructor of literature, I find that aligning and overlapping multiple narrative genres can help my students see that novels have been representing anthropogenic climate change, whether explicitly or unconsciously, for much longer than we might suspect, and that reading noncontemporary literature from a different and potentially much larger perspective links it to our present crisis in compelling ways.

As an example, here I will discuss how I teach Virginia Woolf's *Orlando: A Biography* alongside the Orbis hypothesis, a theory introduced in Simon Lewis and Mark Maslin's essay "Defining the Anthropocene." Readers familiar with these texts will recognize that both, in different ways, tell climatological origin stories that focus on the year 1610. Woolf's experimental novel follows the unnaturally long development of a protagonist whose life begins in the Elizabethan court but who, after more than three hundred years, is just approaching middle age; who changes from man to woman midway through the tale; and who experiences "time passing" over multiple, seemingly incommensurable scales—ranging from the Dallowayan intensity of the single day to the superhuman time spans of the *longue durée*. Lewis and Maslin's essay, meanwhile, follows a different protagonist: an obscure Anthropos whose path can be plotted along a graph of atmospheric carbon dioxide (fig. 2c, p. 174). In both texts, the early 1600s represent a choke point in deep time, a moment that inaugurates fundamental changes not only to geophysical systems but also to the atmosphere that shapes the life of political and cultural forms. In what follows I identify some key moments in *Orlando* that guide our in-class discussion before turning to an explanation of how this novel might be read alongside historical, scientific, and critical sources.

Climate Aesthetics

Although Woolf tends to be celebrated for her commitment to the small scale—for her injunction to "not take it for granted that life exists more fully in what is commonly thought big than what is commonly thought small"—she was also keenly interested in representing "immensities" like national identity, literary history, and deep time, especially by means of popular genres (Woolf, *Common Reader* 150). Eschewing both realism and the granular intricacy of high modernist techniques, *Orlando* experiments instead with a range of generic modes, including romance, biography, and the historical novel. Woolf shifts between these genres and focal lengths, braiding them into a narrative that is simultaneously truthful and fantastic, factual and fictional. We see this scalar shifting at work in the novel's opening section, which lays the foundation of a more or less traditional bildungsroman structure. We are introduced to the titular protagonist, a sixteen-year-old nobleman whose progression into maturity depends on securing the future of his bloodline, title, and estate by marrying. Having already courted two other women, Orlando seems to have found his ideal match in the third. The lady in question is highborn, well-mannered, and, like Orlando, fond of dogs. "In short, she would have made a perfect wife for such a nobleman as Orlando," the biographer tells us, "and matters had gone so far that the lawyers on both sides were busy with covenants, jointures, settlements, messuages, tenements, and whatever is needed before one great fortune can mate with another when, with the suddenness and severity that then marked the English climate, came the Great Frost" (25). Since Orlando's fictional biography has hardly begun, we are prepared for the change of fortune signaled by the conditional perfect tense, "would have." However, the event that arrives at the end of this sentence to disrupt the byzantine legal process of joining two embodied fortunes can hardly be expected. It appears with such "suddenness and severity" that the biographer must break off to report the astonishing scenes that accompanied the unprecedented winter of 1608:

> The Great Frost was, historians tell us, the most severe that has ever visited these islands. Birds froze in mid-air and fell like stones to the ground. . . . The mortality among sheep and cattle was enormous.

> Corpses froze and could not be drawn from the sheets. It was no un-
> common sight to come upon a whole herd of swine frozen immovable
> upon the road. (25)

This weather event dramatically expands our horizon of expectations, caus-
ing Orlando's character development to stall as the narrator moves into a
horizontal survey of historical accounts and anecdotes. The bureaucratic
transactions of the English aristocracy are suddenly swept away by a force
that seems to have neither regard for nor awareness of human affairs. A
multiscalar approach thus becomes essential to the way the novel handles
its historical content, particularly its representations of climate change.

Integrating these sudden climatic changes with Orlando's personal history causes the concept of character itself to expand dramatically, and at
certain points the biographer will attempt to interpret each of Orlando's
attitudes, opinions, and actions through this totalizing concept. This occurs, for example, in a moment when Orlando arouses Queen Elizabeth's
fury by kissing another woman:

> It was Orlando's fault perhaps; yet, after all, are we to blame Orlando?
> The age was the Elizabethan; their morals were not ours; nor their po-
> ets; nor their climate; nor their vegetables even. Everything was differ-
> ent. The weather itself, the heat and cold of summer and winter, was,
> we may believe, of another temper altogether. . . . Translating this to
> the spiritual regions as their wont is, the poets sang beautifully how
> roses fade and petals fall. . . . Violence was all. The flower bloomed and
> faded. The sun rose and sank. The lover loved and went. And what the
> poets said in rhyme, the young translated into practice. Girls were roses,
> and their seasons were short as the flowers'. Plucked they must be be-
> fore nightfall; for the day was brief and the day was all. Thus, if Or-
> lando followed the leading of the climate, of the poets, of the age it-
> self . . . we can scarcely bring ourselves to blame him. (20–21)

This passage confronts us with the startling suggestion that to understand
and perhaps to rationalize the young nobleman's behavior we must look
at his situation climatologically. It posits a strange form of determinism:
was the Elizabethan climate somehow ultimately responsible for Orlando's
alleged transgression?

We must perform vast leaps of logic to reach this conclusion, leaps that
are extended as the novel shifts forward in time. Whereas the Elizabethan
era was marked by polar extremes, Woolf's novel famously describes the
Victorian world in terms of what Gillian Beer calls a "lush and menac-
ing superfecundity" (114). Its milder weather is characterized by a soft,

pillowy atmosphere, dominated by a "great cloud which hung, not only over London, but over the whole of the British Isles on the first day of the nineteenth century." Smudging Britain's contours and contrasts into a Turneresque blur, "it was buffeted about constantly by blustering gales, long enough to have extraordinary consequences upon those who lived beneath its shadow. A change seemed to have come over the climate of England" (Woolf, *Orlando* 166). This change of climate alters the aesthetic qualities of the age itself, casting everything in literally a different light. Scaling out from the domestic to the sociological, the narrator tells us, "The life of the average woman was a succession of childbirths. She married at nineteen and had fifteen or eighteen children by the time she was thirty; for twins abounded. Thus the British Empire came into existence" (168). The fanciful suggestion here is that Britain's colonial territories were conquered as a means of accommodating its surplus population, a surplus that was brought into being, ultimately, by a dramatic change of weather.

The Long View

These are clearly hyperbolic claims, carried to conclusions that seem intentionally absurd. In my seminar, however, I assign material that challenges us to reconsider their legitimacy—to ask whether *Orlando*'s treatment of climate change as the motivating force of a much earlier history might not be as far-fetched as it first appears. Woolf's novel relies on historical records that attest to widespread changes that occurred during an especially acute phase of cold weather (circa 1570–1680) within what has come to be called the Little Ice Age. In class we read one of Woolf's primary sources, "The Great Frost: Cold Doings in London." This tract, first published in January 1608 and reprinted in *An English Garner*, where Woolf discovered it, relates the terrible effects of that famous winter. The narrative unfolds as a dialogue between a citizen of London and a yeoman from the countryside. The Londoner describes "winter castles of ice" that lay piled "against the arches of the Bridge"; he notes that the Thames, frozen solid, has become a highway upon which pedestrians freely cross from the north to south banks "while others play at football" (82, 84). This jovial mood and merriment, however, belies the hardships of a city "cut off from all commerce," beset by an "unconscionable and unmerciful raising of the prices of fuel" (87). The countryman, meanwhile, relates that "[i]t goes as hard with us as it doth with you. You cry out here, you are undone for coals; and we complain, we shall die for want of wood" (88).

The local details preserved in these primary sources can help us sketch the outlines of much larger events that have only recently (and decades after Woolf's writing) begun to disclose the true scale of their causes and effects. In England, the climate of the Little Ice Age led to year upon year of crop failures. Food shortages led to mass starvation, which in turn led to political unrest. Uprisings such as the Midland Revolt of 1607 might, in this sense, be interpreted as direct impacts of severe climate change. In a wide-ranging historical survey, Geoffrey Parker dramatically extends the range of these impacts, linking them to the Scottish Revolution in 1637, the English Great Rebellion in 1642, and a series of major revolts and revolutions that occurred across Europe, the Americas, Asia, and Africa.

Yet this climatological backstory can be traced even further into the past if we read it in the context of the Orbis hypothesis. Lewis and Maslin's findings indicate that the severe temperature shifts associated with the Little Ice Age may have had anthropogenic causes, starting with the "arrival of Europeans in the Caribbean in 1492, and subsequent annexing of the Americas," which "led to the largest human population replacement in the past 13,000 years" (174). This event had such extensive ecological ramifications, the authors argue, that it qualifies as a truly global terraformation:

> Besides permanently and dramatically altering the diet of almost all of humanity, the arrival of Europeans in the Americas also led to a large decline in human numbers [from] a total of 54 million people in the Americas in 1492, with recent population modelling estimates of 61 million people . . . to a minimum of about 6 million people by 1650 via exposure to diseases carried by Europeans, plus war, enslavement and famine. The accompanying near-cessation of farming and reduction in fire use resulted in the regeneration of over 50 million hectares of forest, woody savanna and grassland. . . . The approximate magnitude and timing of carbon sequestration suggest that this event . . . is the most prominent feature, in terms of both rate of change and magnitude, in pre-industrial atmospheric CO_2 records over the past 2,000 years. (174–75)

Geological evidence indicates, in other words, that as much as ninety percent of the existing human population of the Western Hemisphere died within about 150 years of European contact and conquest and that the resulting reforestation of those emptied continents was sufficient to drop the temperature of the planet for a period that lasted "about 330 years" (Matthews and Briffa 17).

When discussing the Orbis hypothesis as a potential golden spike for the Anthropocene, I recall Yusoff's critique that using familiar terms like "the discovery of the new world" or the Columbian Exchange "covers over the friction of a less smooth, more corporeal set of racialized violences" (29). Deploying "the language of exchange," Yusoff points out, encourages the assumption that "something was given rather than just taken," when in truth, European invaders gave only new diseases, suffering, and death (29). From this perspective *Orlando* might be read not just as a novel about climate change but as a work that records the indelible link between climate change and colonial violence. In this way, *Orlando* allows us to grasp how "the unsettling of narrative and historical time frames wrought by climate change," as Jesse Oak Taylor persuasively argues, "affects not merely the way we read texts in the context of their futurity but also the retrospective truth of any interpretive encounter."

Woolf's conflation of climate and national identity in *Orlando* takes on a new resonance when we encounter the freezing and thawing of the British Isles—and, by extension, of the planet—as its organizing chronotope. Pulling at the novel's unusually long narrative threads, we find the Great Frost entangled with Britain's subsequent industrialization, since the severely cold winters of the Little Ice Age contributed to the mass consumption of coal, which in turn lit the spark of an engine that has been warming the planet ever since. From *Orlando*'s violent opening scene of a boy using the preserved, severed head of an African man for fencing practice to its lavish depictions of frost fairs on the Thames to its sweeping visions of Britain's industrialization, the novel testifies to the charge that colonial regimes were directly responsible for shaping severe weather events that have punctuated history from the early modern period to the present.

Works Cited

Beer, Gillian. *Darwin's Plots: Evolutionary Narrative in Darwin, George Eliot and Nineteenth-Century Fiction*. Cambridge UP, 2009.

"The Great Frost: Cold Doings in London, Except It Be at the Lottery." *An English Garner: Ingatherings from Our History and Literature*, edited by Edward Arber, vol. 1, Westminster, 1897, pp. 77–99.

Lewis, Simon L., and Mark A. Maslin. "Defining the Anthropocene." *Nature*, vol. 519, no. 7542, 11 Mar. 2015, pp. 171–80.

Luciano, Dana. "The Inhuman Anthropocene." *Avidly*, 22 Mar. 2015, avidly .lareviewofbooks.org/2015/03/22/the-inhuman-anthropocene/.

Matthews, John A., and Keith R. Briffa. "The 'Little Ice Age': Re-evaluation of an Evolving Concept." *Geografiska Annaler: Series A, Physical Geography,* vol. 87, no. 1, 2005, pp. 17–36.

Parker, Geoffrey. *Global Crisis: War, Climate Change and Catastrophe in the Seventeenth Century.* Yale UP, 2013.

Taylor, Jesse Oak. *The Sky of Our Manufacture: The London Fog in British Fiction from Dickens to Woolf.* U of Virginia P, Kindle ed., 2016.

Woolf, Virginia. *The Common Reader.* Edited by Andrew McNeillie, vol. 1, Vintage, 2007.

———. *Orlando: A Biography.* 1928. Harcourt, 2006.

Yusoff, Kathryn. *A Billion Black Anthropocenes or None.* U of Minnesota P, 2018.

Part IV

Courses and Interdisciplinarity

Hannah Kroonblawd

It's the End of the World As We Know It: Utilizing Interdisciplinarity to Teach Anthropocene Literature

In November 2019, I received my teaching assignment for the upcoming spring semester: two sections of an undergraduate interdisciplinary literature course. It is a special topics course, and, given my research interests and the work I'd been doing for my comprehensive exams, I decided to design a reading list centered on survival, Anthropocene, and apocalypse and to call the course It's the End of the World As We Know It. If I'd known then what was ahead—that the coming spring would be a tumultuous one physically, emotionally, spiritually, and virtually—would I have changed the course's theme? I don't think so, because I'd like to think that the work students and I did in the early weeks of 2020—the conversations we had, the journals we wrote, the books we read, and the films we watched—helped prepare us for the months ahead.

That fall, as I began to brainstorm the goals for the course, alongside compiling a range of textual and audiovisual materials, the idea web stretched from ecocriticism and activism to economics and industry. The Anthropocene, literally "the human epoch," has gained traction over recent years as a label for ecological, geological, industrial, and economic changes that have occurred since the mid–nineteenth century (see Farrier 4–5). In some ways, a term like *Anthropocene* can distill contemporary existence into an oversimplification: human existence is the primary power on Earth, for better or for worse. To address issues with this kind of generalization, some scholars have proposed terms like *Capitalocene* and *Necrocene* (see J. Moore 237; Casid 30). But *Anthropocene* has its uses, and its implications give a name to the confluence of power and subjugation, whether political, fiscal, racial, environmental, social, gendered, abled, or religious. Climate change is never just climate change alone.

The Anthropocene asks us—as teachers and scholars, readers and writers, whether within or without the academy—to consider what particular spheres of influence and positionality exist in the face of larger structures of power, what the history of such structures indicates about their continued existence, and what expectations might be held for our personal and collective futures. Literature asks the same. Inside the college classroom, the intersection of contemporary text and critical thought provides a location for complicating Anthropocene futures, calling for resistance, expressing horror, and conveying hope. Approaching the Anthropocene, and climate change, within the literature classroom allows students to interface with still-developing, uncodified, dynamic theoretical frameworks. Questions of academic authority, disciplinary canon, and situated ethics can deepen student interaction with text, theory, and the acts of reading and writing—no matter the genre at work. And the Anthropocene is lived out beyond the classroom. It is our day-to-day, which became more and more evident to our interdisciplinary classroom between January and May of 2020.

In preparing to teach an interdisciplinary literature class, which fulfills a general education humanities requirement and is often taken by first- and second-year students at my large public university, it became important for me not only to define the theme of the course but to better understand interdisciplinarity itself. A bit like *Anthropocene*, the term *interdisciplinarity* encompasses a variety of definitions. Rick Szostak, a professor of economics at the University of Alberta, lists areas of consensus among university approaches: complex problems or questions whose scope is beyond a single discipline, use of specialized research from or of individual disciplines, a multiplicity of methods, and the affordances that a comprehensive and integrative perspective can bring to a particular issue. Clinton Golding, of the University of Otago, writes that interdisciplinarity can be approached through disciplinary breadth or disciplinary depth—but emphasizes that depth is not a necessary prerequisite for interdisciplinary learning: "The hallmark of an interdisciplinarian is meta-disciplinary understandings and skills, and the development of these is the main objective of an interdisciplinary subject. Students in these subjects should learn how to access, understand, employ, and synthesize the expertise from various disciplines" (5). Much of the work of an interdisciplinary perspective is akin to the way genre studies approaches are taught within first-year composition classes: identifying and moving through different ways of knowing.

Interdisciplinarity is well suited to tackling the questions of the Anthropocene—questions of scale, survival, sustainability, and speculation. It is also a useful approach for a survey-level general education class—which is sometimes the only literature course a student will take during college—and provides a singular opportunity for engagement beyond composition. My students themselves were members of an interdisciplinary cohort: in a single section of thirty students, majors ranged from nursing to marketing, elementary education to kinesiology. I had one student who was an English major, so she and I were the two who brought the English discipline's way of knowing. Part of my task for this class became working with students toward the recognition of their disciplinary knowledge and the value of sharing that knowledge with and alongside others.

In choosing texts for the course, I wanted to incorporate the typical slate of literary genres—short stories, poetry, and a novel—as well as a few others: film, theater, and graphic novelization. The finalized text list included

> *Severance*, by Ling Ma: a novel about a young woman working in New York City as an unprecedented, memory-altering pandemic sweeps across the world;
>
> *Extinction Events*, by Liz Breazeale: a collection of short stories about cataclysms both personal and collective, ecological and sociological;
>
> *Lake Michigan*, by Daniel Borzutzky: a sequence of poems that illuminate the terror within Chicago's incarceration system;
>
> *Deaf Republic*, by Ilya Kaminsky: a narrative poetry sequence telling the story of a town that goes silent in the face of violent occupation;
>
> *V for Vendetta*, by Alan Moore and David Lloyd: a graphic novel (first serialized in 1982) depicting a totalitarian British government opposed by a man called V;
>
> *Hadestown*, by Anaïs Mitchell and Rachel Chavkin: a musical adaptation of the Orpheus and Eurydice myth, set in a weather-ravaged

> land where an industrial underworld's labor system seems the
> only possible future; and
>
> *Hunt for the Wilderpeople*, directed by Taika Waititi: a film following
> the misadventures of a boy and his foster father as they go on the
> run through the New Zealand bush.

Not every text we examined is climate-change-oriented or blatantly Anthropocentric, but each provides a unique perspective on what it means to live in a human epoch. Some of these narratives are noticeably connected to the ecological concerns of the Anthropocene; others examine capitalism, racism, and government control. The documentary film *Anthropocene: The Human Epoch* was a spur-of-the-moment and rewarding addition to course material that diverged from the narrative-driven structure of other texts. As we moved to online learning midway through the term, other film selections also shifted—initially I'd planned for *Wall-E* and *Children of Men* but needed a film all students could access. More recently, I've taught the graphic novel adaptation of Octavia E. Butler's *Parable of the Sower* and have added it to the syllabus for future semesters.

One aim in text selection was, like one approach to interdisciplinarity, breadth. I hoped students would be able to articulate other narrative encounters from their day-to-day—in the music they listen to, the TV shows they watch, the social media they scroll through, and the ways they tell one another their own stories. I wanted students to practice clear expression of how worldview and context influence those narratives, whether through image or song or written word, and in particular how those narratives speak to human conceptions of time and progress. Many students had expected to spend the semester reading and writing; it was a surprise to some that they could use those same tools, coupled with their own disciplinary expertise, to analyze how comedy-drama films from New Zealand and Tony Award–winning musicals speak to deep time and geologic records. The element of surprise is important to me in course design—it adds excitement when students can't quite see at first how each piece fits into the whole.

Introducing interdisciplinarity also employs fun and imaginative work at the start of the term. On the first day of class, I asked students to reflect on this question: "When faced with disaster or catastrophe, how would someone in your field of study respond?" Students considered disciplinary priorities, areas of expertise, and, maybe most revelatory, where their discipline might fall short. They then joined groups with those from other fields, discussing assumptions and stereotypes held about different disciplines, comparing their differing disaster responses, and noting where strengths and weaknesses were in balance. Beginning class in this way gave students some breathing room before we began a vocabulary-intensive introduction to the Anthropocene. Verbalizing personal and disciplinary knowledge and capacity spun the class in the opposite direction from the way courses usually begin. Instead of starting with "what you will learn," we began with "what you know and can share and what others are able to share with you."

The second week focused on the Anthropocene. We asked questions about the Great Acceleration of human activity over the past seventy-five years, compared scarcity with abundance, and journaled about global, local, and personal values. In most classes I teach, we discuss the three primary questions any worldview, theoretical approach, or individual reader might try to answer: What is good? What exists? What is possible? In an intentionally interdisciplinary course, we are able to lay out how different disciplines answer these three questions, and in a course themed around the Anthropocene, we push these questions to their extremes. What is good about the twenty-first century? What dangers exist? What kind of future is possible?

Because this class is about breadth, I provided a crash-course introduction to literary theory and its history. I didn't expect students to become literary critics overnight, but I wanted them to be able to identify different ways a text has been or could be read—what it means if a reader focuses on the text itself, authorial context, or audience reception. Reader-response theory, postmodernism, and new historicism appear and are useful as disciplinary terminology, but my emphasis was on their implications rather than their intricacies. As I am a creative writer, I also gave students some discipline-specific knowledge regarding literary publishing—how a book moves from author to audience—and the power social media platforms and book reviews have to affect book reception.

Once it was clear how everyone—not just myself as the instructor—brought unique disciplinary knowledge into the classroom, how the theme connected with questions we would ask of the texts, and how literature as a discipline gathers its knowledge, we dove into our first book, Ma's *Severance*.

When I look back at my notes from the first few weeks of class, I can track the events that were occurring around us. In early January, it was the wildfires tearing across Australia, where one student had studied abroad the semester before; coronavirus doesn't explicitly enter my teaching notes until 28 January. A different pandemic had come up in class days before, since *Severance* is, among many things, a pandemic story. It was as if we were watching the events of *Severance* play out in real time: the move from school and workplace into the home, the designation of "essential" workers, the flight from large urban centers, the rise of conspiracy theories and political rhetoric coupled with public health issues, the disparities illuminated by community spread and the ability or inability to quarantine. We tracked where fact differed from fiction and where novelization mirrored news stories. Eventually, students journaled about their semester-from-home experiences, evoking the photo blog *Severance*'s narrator uses to document her solitude in an emptied-out city. "Are you psychic?" more than one student asked me in March. "Did you know this would happen?" Teaching disaster-oriented, climate change, or Anthropocene literature means grappling with this very question: How are writers so attuned to what could occur? Or, put another way, Why does speculative literature so often become prophetic?

Breazeale's *Extinction Events* provided opportunity for similar analysis, given that the third story in the collection, "Survival in the Plague Years," moves across history from one epidemic to another. The short stories also allowed us to turn our attention toward environmental disasters, some slow-moving, others instantaneous. Students utilized their disciplinary awareness as they read: elementary education students discussed the portrayal of childhood; sociology and psychology students described family dynamics and trauma; art students focused on description and imagery; computer science students noticed technological threads. One student was preparing for a spring break geology trip; her favorite story was a fossil-focused family drama titled "Devil's Tooth Museum." The Anthropocene hinges on scale, and so do these stories—students were quick to note how time shifts its scope, how the short story differs from the novel, and how stories change depending on how and when and why we tell them.

It wasn't always easy for students to bring their own discipline into the literature classroom or for me to consistently engage with such a disparate range of interests. There were some slower, quieter class days, especially as concern grew about the unknowns of our own pandemic. At the beginning of the semester, sharing was a challenge—perhaps because students are frequently trained to enter class with an expectation of receiving rather than sharing knowledge. To acclimate students to a recognition of their own systems of knowledge, I used a pattern of in-class free-writes, out-of-class journals, similar-discipline check-in groups, and interdisciplinary focus groups. When grouped with peers in similar majors or programs of study, students leveraged and deepened disciplinary expertise, reviewing what they were learning in other classes and how their discipline spoke back to the literature we were reading. When working within interdisciplinary groups, students turned stories into case studies and case studies into stories. Those studying engineering explained the logistics of power grid failure, biology majors brought in statistics from conservation classes, and political science students turned to news blurbs about environmental legislation. I had to be explicit and repetitive in requesting that students share what they know, what their discipline knows, and what they would like to know from others. Groups don't work if only one person is working. "This is how we problem-solve—together," I told them. "This is how we learn from narrative and discipline and one another." It was unusually hard work that only became harder in an unusual semester.

Each unit—fiction, poetry, and multimedia narratives—culminated in a hybrid analysis, where students had different options for responding to readings. Some focused on how their own discipline would answer the questions raised within different texts, others on using vocabulary of the Anthropocene in connection with conventions of the literary genres they had read. I reconfigured the final project to accommodate a very difficult end of the term, asking students to revisit the worldview questions from the beginning of the semester in the light of what the COVID-19 pandemic had brought to their lives. Students wrote about quarantine, about toilet paper shortages, about babysitting their siblings or renegotiating boundaries with their parents; but they also wrote about isolation, fear, and anxiety. Some wrote about news analyses and misinformation, some about supply chains and their essential-worker jobs at grocery stores, and others about navigating the pandemic as an immunocompromised person. Many wrote about religion, community, sacrifice, and hope.

In the end, it wasn't the course title or theme that caused students to push toward the Anthropocene and what it implies about humanity's movement through and across and upon the world. Interdisciplinarity became real as the very stories we read became lived reality—the reality that we don't have to face problems alone; that we hold different kinds of valid, valuable knowledge—and that together we hold even more; that having knowledge is different than using it; and that reading a story can teach us how to live.

This is the lesson of Anthropocene literature, of the literature of climate change, of activist literature. It is also the lesson of the interdisciplinary classroom. I know I learned just as much as my students did as I listened to their questions, read their reflections, and watched them problem-solve. I saw their resiliency. I learned to be more specific in asking them to share their knowledge, which meant I learned more and more about them as individuals. I recognized that there were moments—more often than not—when I could take a step back and let them lead, because they knew things, so many things, that I did not. Understanding the Anthropocene and its implications for the future means understanding that the way forward is not singular but collective; interdisciplinary learning is essential to harnessing hope by means of the knowledge that we—students, teachers, readers, writers—have to share with one another.

Works Cited

Anthropocene: The Human Epoch. Directed by Jennifer Baichwal et al., Mercury Films, 2019.

Borzutzky, Daniel. *Lake Michigan.* U of Pittsburgh P, 2018.

Breazeale, Liz. *Extinction Events: Stories.* U of Nebraska P, 2019.

Butler, Octavia E. *Parable of the Sower: A Graphic Novel Adaptation.* Adaption by John Jennings and Damian Duffy, Abrams, 2000.

Casid, Jill H. "Doing Things with Being Undone." *Journal of Visual Culture,* vol. 18, no. 1, Apr. 2019, pp. 30–52.

Farrier, David. *Anthropocene Poetics: Deep Time, Sacrifice Zones, and Extinction.* U of Minnesota P, 2019.

Golding, Clinton. *Integrating the Disciplines: Successful Interdisciplinary Subjects.* Centre for the Study of Higher Education, 2009.

Hadestown. Written by Anaïs Mitchell and directed by Rachel Chavkin, original Broadway cast recording, Sing It Again, 2019.

Hunt for the Wilderpeople. Directed by Taika Waititi, Piki Films, 2016.

Kaminsky, Ilya. *Deaf Republic.* Graywolf Press, 2019.

Ma, Ling. *Severance.* Picador, 2019.

Moore, Alan, and David Lloyd. *V for Vendetta.* DC Comics, 2020.

Moore, Jason W. "The Capitalocene Part II: Accumulation by Appropriation and the Centrality of Unpaid Work/Energy," *The Journal of Peasant Studies,* vol. 45, no. 2, 2018, pp. 237–79.

Szostak, Rick. "Defining 'Interdisciplinary.'" *Rick Szostak,* 2015, sites.google .com/a/ualberta.ca/rick-szostak/research/about-interdisciplinarity/ definitions/defining-interdisciplinary.

Patrick Whitmarsh

"It Will Take Years for the Picture to Emerge": Interdisciplinarity, Intermedia Strategies, and Climate Narratives

Those familiar with the scientific and political literature on climate change know that it has often been described as a "wicked problem" (if not a "super wicked problem"; see Incropera). Despite its colloquial overtones, the term announces a set of characteristics that make solving the problem of climate change notoriously difficult—the most distinctive, perhaps, being irreducibility. Addressing one aspect of the issue impacts all others, yet all demand action. For this reason, the wicked problem of climate change exceeds the scope of any single discipline. Those who specialize in marine ecosystems are adept at explaining the loss of coral reef habitats but are not necessarily adept when it comes to deconstructing the advertising campaigns of fossil fuel conglomerates. Those of us who practice literary criticism are likely more attuned to the cultural narratives concerning fossil fuels and environmental responsibility but less attuned to the carbon emissions of the United States military or to the stratigraphic research that looks for evidence of the Anthropocene. In a culture structured around the central feature of professionalization, universities have had to acknowledge the interdisciplinary demands of the climate crisis. A solution requires not only international cooperation but also interdepartmental cooperation between scientists, humanists, politicians, economists, and others.

For those of us in English and other literature-based departments, teaching the literature of climate change is also a kind of wicked problem, requiring us to deal with one major challenge of climatological complexity: namely, that the climate crisis we currently inhabit is born of long, gradual change over time. Given literary narrative's attention to time and temporality, we might assume the capacity of literature to engage actionably with the long problem of climate change, and yet narrative fiction has often encountered certain limits on this front. Even that most temporally focused of literary modes, the historical novel, runs up against the dilemma of telescoping human history within the deeper historical spiral of geological time. On the one hand, this dilemma is itself a historical phenomenon. Only recently—within the last seventy-five years or so—have humanists realized the necessity of studying human and geophysical history simultaneously and the historiographic implications of doing so. On the other hand, however, the novel appears to encounter certain restraints at the level of form: "Within the mansion of serious fiction," writes Amitav Ghosh, "no one will speak of how the continents were created; nor will they refer to the passage of thousands of years: connections and events on this scale appear not just unlikely but absurd within the delimited horizon of a novel" (61). For whatever reason, "serious fiction" encounters difficulty when formalizing change on a planetary scale.

The implicit sense of Ghosh's comment is that he wants to see fiction depicting climate change realistically. For Ghosh, the speculative futures in works like Kim Stanley Robinson's *New York 2140* and the surreal expressionism of Jeff VanderMeer's *Southern Reach* trilogy fail to answer the call to render climate change as it is happening now, as we see it in the world around us. Of course, one problem with this clarion call for "serious" climate fiction is that not all of us see climate change in the world around us; or, rather, we do not realize that we see it. We may see weather, construction, cars, concrete, infrastructure, trees, flowers, and birds whose presence and behavior are part of climate change, but they may not speak to us as climate change. Ghosh insists on a kind of literature that would reveal the intricate web of climate and ecology that links these elements.

To some who have responded to Ghosh, his lament tastes a bit sour (see, for example, Marshall). It can be argued that in many respects, it is not a literature of realism that we need but a literature of derealization, a literature that shows us we are now living in science fiction, as Robinson has quipped. Yet the problem remains of connecting these strange and nonrealist literary endeavors to our students' everyday lives—of persuading them that their experience *is* that of climate change, even if they do not always see it that way. To what extent can we mobilize the significance and urgency of climate change within the intimate and personal setting of the classroom? How do we convey the slowness of global warming when we only meet with our students for a few hours at most each week? The bleak irony is that as we move further into the twenty-first century, the slowness of climate change has begun to pick up speed. Scientists, activists, politicians, and the media are paying more attention to shifting weather and migration patterns, the seasonal windows of hurricanes and wildfires, the intensity of heat waves, and the decreasing amount of polar ice. Even with these alterations, however, climate change remains an all-too-often unrecognized phenomenon.

I therefore highlight here an exemplary network of materials that invites instructors and students to bridge the gap between the realist and the speculative, between our daily lives and the slowness of climate change. In my own experience as a teaching assistant for Boston University's Interdisciplinary Perspectives on Global Challenges: Climate Change, a lecturer for Fighting the Climate Crisis: From Earth Day to Extinction Rebellion (a course I designed and taught for Harvard University's History and Literature program), and a visiting assistant professor of environmental studies at Wofford College, these sources have proven to be both engaging and effective in drawing students' attention to the visceral reality of climate change. Combined with an approach that pushes literacy across disciplines, the following texts offer students opportunities to grapple with not only the science of climate change but also its cultural meanings and narratives.

I have learned that students often respond well to a bit of theoretical framing before being asked to dive into primary sources and cultural artifacts. Focusing on the concept of slowness, Rob Nixon's *Slow Violence and the Environmentalism of the Poor* offers a helpful analogue for thinking through the difficulty of climate temporality. Although its main concern is not climate itself but rather the prolonged suffering brought about by resource extraction and settler colonialism, Nixon's book conceptualizes slowness in a way that enables students to think concretely about the glacial temporalities of the planet. "In the long arc between the emergence of slow violence and its delayed effects," Nixon writes, "both the causes and the memory of catastrophe readily fade from view as the casualties incurred typically pass untallied and unremembered" (8–9). Likewise, the long arc of climate change can make it difficult to associate events today with an industrial history that began centuries ago or to imagine a future in which our lives are radically altered. Planetary timescales do not lend themselves easily to human perception. It requires work to see the historical tendrils that wind through centuries of development. *Slow Violence and the Environmentalism of the Poor* offers a clear and cogent idea—that of slow violence itself—that empowers students as they try to make sense of the dynamic between climate and culture.

A grasp of slowness sheds light on the temporal incongruities that crop up in the literature of climate change. In Richard Powers's *The Overstory*, such incongruities appear as one of the novel's central themes. "To be human is to confuse a satisfying story with a meaningful one," one of its characters muses, "and to mistake life for something huge with two legs. No: life is mobilized on a vastly larger scale, and the world is failing precisely because no novel can make the contest for the *world* seem as compelling as the struggles between a few lost people" (383). The takeaway of Powers's metatextual description is that narrative, as a concept, need not conform to realist standards of time but can be adapted to tell nonhuman stories. "It will take years for the picture to emerge," *The Overstory*'s Patricia Westerford explains, describing the process by which tree roots grow to comprise a network (142). Such moments are pedagogical lessons if we are willing to learn from them. *The Overstory* is not only a fiction about human lives brought together (and torn apart) by trees; it is an instruction manual in how to reframe narratives at the scale of climate and the environment and how to perceive the ways these scales impact our daily lives.

In both its formal concerns and its thematic ones, Powers's novel enacts the intersections between social and natural systems as well as the mutual benefits shared between them: "[T]rees want something from us," Westerford says, "just as we've always wanted things from them" (454). In the field of ecology, this relationship is known as "ecosystem services," or benefits "derived from nature without the interference of a market" (Boumans et al. 32). Roelof Boumans, Joe Roman, Irit Altman, and Les Kaufman present a model they call MIMES (the Multiscale Integrated Model of Ecosystem Services) to capture the exchanges between human and natural systems—exchanges we can see playing out in the narrative structure of *The Overstory*. As the MIMES model emphasizes, the recent ecological paradigm recognizes "that each subsystem is characterized via its embeddings within the other," meaning that natural ecosystems can no longer be viewed as external to human social systems and that each determines the state of the others (30). *The Overstory* assumes this embedment from its opening pages, constructing a narrative in which meaning derives from humanity's entanglement with nonhuman ecosystems.

This kind of complementary assessment informs the interdisciplinary methods that enhance the ways we teach the literature of climate change. The scientific model of MIMES enables students to identify key passages in *The Overstory* and to make insightful connections between the scientific findings and Powers's narrative strategies. For courses whose instructors find themselves pressed for time, shorter works can achieve similar ends. For example, the Boston University (BU) course assigned Primo Levi's flight of fancy "Carbon" as one of the first readings in the semester, even before teaching the basics of the carbon cycle, human agricultural and industrial history, or the warming potential of greenhouse gases. In this short story, the narrator follows a carbon atom through a series of material encounters on Earth, emphasizing both its temporal longevity and spatial granularity: "Its existence," Levi writes, "whose monotony cannot be thought of without horror, is a pitiless alternation of hots and colds, that is, of oscillations (always of equal frequency) a trifle more restricted and a trifle more ample: an imprisonment, for this potentially living personage, worthy of the Catholic Hell" (226). The story expresses the unimaginability of the atomic scale, yet this unimaginability is draped in affect. It gives the atom a human face, so to speak, and in more ways than one; by the end of Levi's tale, readers learn that the carbon atom has entered one of the narrator's brain cells and is involved in the very composition of "Carbon."

"Carbon" and *The Overstory* offer creative demonstrations of the entanglement between human and geological agents, effectively realizing Dipesh Chakrabarty's reminder that humans "can become a planetary geological agent only historically and collectively, that is, when we have reached numbers and invented technologies that are on a scale large enough to have an impact on the planet itself" (31). The first chapter of *The Climate of History in a Planetary Age* (23–94) pairs well with sources that demonstrate or perform geological agency. Such texts enable students to grasp the critical implications of Chakrabarty's argument, including the merging of human and geological calendars, and the scale Chakrabarty employs likewise informs the presentation of the companion texts. In my course, I introduce students to the premise of geological agency through a very explicit example: the earthmoving tactics of "peaceful nuclear explosions" in the mid–twentieth century, particularly those of Project Plowshare. As an assignment, students watch *Plowshare*, a promotional video from the

Atomic Energy Commission advocating for the development of nuclear strategies for engineering shipping and trade routes. Students often find the imagery and propagandistic tone of *Plowshare* compelling, which makes for productive classroom discussion. I pair the video with readings on the ecological impact of radioactive fallout, such as Rebecca Solnit's *Savage Dreams: A Journey into the Hidden Wars of the American West*, which declares, "Much of the fallout of the fifties is still working its way through the plants and bodies of the living things in the places it fell, or lying on the ground waiting to circulate through these living systems" (96). Solnit offers a critical perspective on the US nuclear program and illuminates the deeper and less seen ways that atomic testing seeps into regional ecologies, affecting organisms for generations to come. Through this textual pairing, students understand that geological agency is not simply the practice of moving earth to create new canals and waterways but also the irradiation of water and soil. The violent spectacle of detonation stands in contradistinction to the slow invisibility of fallout, providing students with another opportunity to wrestle with the dilemma of intuiting climate slowness.

Whereas *Plowshare* underscores the dramatic alteration of landscapes and ecosystems in an instant, other visual sources, such as Jeff Orlowski's documentary *Chasing Coral*, are particularly good at illuminating ecological transformation over extended periods of time. Orlowski's film juxtaposes archival recordings of thriving, healthy reefs from decades ago with recent images of sick and dead coral. BU students reported being struck by the differences between the old footage and the new, noting that even visiting the reefs firsthand would not provide this comparative opportunity. *Chasing Coral* underscores the stark contrasts visible in images that document change over time, drawing viewers to a scene that is largely invisible in our everyday lives. Orlowski's film pairs remarkably well with Richard McGuire's graphic novel *Here*, which employs the same contrast to a more extreme degree and in a setting that will be more familiar to students. Every page of McGuire's novel offers a view of the same space over the course of planetary time. Although most of the panels are occupied by interior views of a house built on the property, the author intersperses smaller frames embedded on the page, offering glimpses of alternative time periods—some in the deep past and some in the distant future. Setting a camera-like focus on a single spatial perspective, *Here* renders visible the passage of geologic time and is a wonderful complement to scientific discussions of the concept, such as Marcia Bjornerud's magnificent *Timefulness: How Thinking Like a Geologist Can Help Save the World*.

"[My] daily experience as an earthling is enriched by an awareness of the lingering presence of so many previous versions and denizens of this place," Bjornerud writes. "Understanding the reasons for the morphology of a particular landscape is similar to the rush of insight one has upon learning the etymology of an ordinary word" (17–18). The urgency of climate change demands an appreciation for planetary timescales, and assigning McGuire and Bjornerud together conveys their magnitude, grounding students in the sublimity of planetary change.

The literature of climate change is not only fiction or only scientific publications; it is the dynamic that emerges between very different narrative modes. It encourages an interdisciplinary approach comprising close readings and discussion of theoretical analogies, keeping an eye out for the ways that scientific concepts such as MIMES play a role in literary representation. One upshot of this methodology is that it allows instructors to model the very same kind of "wicked problem" that we face with climate change: presenting the literature as a complex system assists students in exploring the ways that alternative perspectives (or disciplinary approaches) highlight discrete elements of a whole. Approaching the materials this way reminds students of their agency—both geological and social. As current inhabitants of this planet, students have both the power and the responsibility to act ethically toward the environment and future generations. The intermedia materials and interdisciplinary demands of climate change literacy give students the intellectual tools to participate in the public discourse and become part of the climate justice movement.

Works Cited

Bjornerud, Marcia. *Timefulness: How Thinking Like a Geologist Can Help Save the World*. Princeton, 2018.

Boumans, Roelof, et al. "The Multiscale Integrated Model of Ecosystem Services (MIMES): Simulating the Interactions of Coupled Human and Natural Systems." *Ecosystem Services*, vol. 12, 2015, pp. 30–41.

Chakrabarty, Dipesh. *The Climate of History in a Planetary Age*. U of Chicago P, 2021.

Chasing Coral. Directed by Jeff Orlowski, Exposure Labs, 2017.

Ghosh, Amitav. *The Great Derangement: Climate Change and the Unthinkable*. U of Chicago P, 2016.

Incropera, Frank P. *Climate Change: A Wicked Problem: Complexity and Uncertainty at the Intersection of Science, Economics, Politics, and Human Behavior*. Cambridge UP, 2016.

Levi, Primo. "Carbon." *The Periodic Table*, by Levi, translated by Raymond Rosenthal, Schocken, 1984, pp. 224–33.

Marshall, Kate. "The Readers of the Future Have Become Shitty Literary Critics." *B2O: The Online Community of the* Boundary 2 *Editorial Collective*, Duke UP, 26 Feb. 2018, www.boundary2.org/2018/02/kate-marshall-the -readers-of-the-future-have-become-shitty-literary-critics/.

McGuire, Richard. *Here.* Pantheon, 2014.

Nixon, Rob. *Slow Violence and the Environmentalism of the Poor.* Harvard UP, 2011.

Plowshare. U.S. Atomic Energy Commission, circa 1961. *Internet Archive,* uploaded 7 July 2012, archive.org/details/0418_Plowshare_09_00_47_00.

Powers, Richard. *The Overstory.* W. W. Norton, 2018.

Robinson, Kim Stanley. "The Realism of Our Times: Kim Stanley Robinson on How Science Fiction Works." Interview by John Plotz. *Public Books*, 23 Sept. 2020, www.publicbooks.org/the-realism-of-our-times-kim-stanley-robinson -on-how-science-fiction-works/.

Solnit, Rebecca. *Savage Dreams: A Journey into the Hidden Wars of the American West.* 1994. U of California P, 2014.

Cynthia Schoolar Williams

Reading the Weather: Teaching the Literature of Climate Change at a Polytechnic University

At Wentworth Institute of Technology in Boston, engagement with literature often emerges within a multidisciplinary setting. Historically, the school has emphasized the built environment, offering degrees in architecture, industrial and interior design, eight fields of engineering, construction and facilities management, and business. We have recently achieved university status, and our School of Sciences and Humanities has launched its first bachelor of science degrees (in applied science and in computer science and society), celebrated milestones in keeping with the experiential, multidisciplinary orientation of the institute. In this energized and evolving setting I've now taught my environmental humanities class, Reading the Weather, twice, first in the fall of 2019 and again, remotely, in the fall of 2020. I believe the course offers a valuable model for teaching the literature of climate change in a polytechnic setting, where a pragmatic student body is preparing for the hands-on work of adaptation and mitigation.

Approach to Content

My first decision in creating the class was not to structure the content chronologically. Rather than trace the narrative of ecocriticism from nature writing in the nineteenth century to the climate fiction of the twenty-first, I organized content in clusters. From the many manifestations of climate disruption, I chose a small set of natural elements and processes that carry a heightened sense of urgency today, and for each of these I assembled materials from a range of disciplines, including and prioritizing literature. Our clusters, which I describe below, covered forests, waterways, and ice, and we began to engage with them in the fourth class of the semester. In the first three sessions, we explored two complex terms that have outsize implications for the literature of climate change: *Anthropocene* and *wilderness*. Of course, I could have chosen many other terms for our initial foray, but these two served my purposes.

I wanted my students to understand the expansive reach of *Anthropocene* far beyond its origins in the field of geology. We read Paul J. Crutzen and Eugene F. Stoermer's "The 'Anthropocene,'" Joshua Howe's "This Is Nature; This Is Un-Nature: Reading the Keeling Curve," and Simon L. Lewis and Mark A. Maslin's "Defining the Anthropocene." We discussed the cascading consequences of each proposed starting point for the geologic age and how our sense of human history must change depending on when we determine the Anthropocene to have begun. (By vote, both classes chose 1784 rather than the International Union of Geological Sciences' 1950.) The second term, *wilderness*, we recognized as being especially dynamic in US literature and culture, and we were aided by two readings: William Cronon's "The Trouble with Wilderness" and Mark Dowie's "The Myth of a Wilderness without Humans." I surveyed the students about their associations with wilderness and played back to them the indeterminacy that prevails over the term's ostensible clarity.

From engaging with these two evocative terms, we realized, first, that language around climate change is rapidly evolving and, second, that many key concepts in this public discourse have a fraught cultural history demanding our attention. Moreover, when faced with a "wicked problem" (first described by Horst Rittel and Melvin Webber as a unique problem for which there is no immediate test of a solution and no opportunity to test solutions through trial and error), disciplines must become ever more mutually informing. Here we took encouragement from Maureen McLane's poem "OK Let's Go," which beautifully reminds us of the need to learn, unlearn, and relearn—a dynamic we would return to throughout the semester.

Having activated a multidisciplinary approach, we engaged with our first cluster: forests. We began with data both historical and current (e.g.,

Kaplan et al.; Marinelli; Robbins). We read Gary Snyder's "Mid-August at Sourdough Mountain Lookout" in the light of forest fires now burning. We paired McLane's poem "Märchen" with popular coverage of Europe's last, endangered fairytale forests (Hester). Especially intriguing were two essays that evoke the narrative of US settlement from the eastern forests to the open prairie: Henry David Thoreau's "Walking" and Alan Bewell's "Jefferson's Thermometer," which draws intricate connections between national identity and climate. The second time I taught the course, I amplified our focus on Indigenous rights and complemented "Jefferson's Thermometer" with Kyle Powys Whyte's "White Allies, Let's Be Honest about Decolonization."

Response papers and discussion boards kept our conversation going, and for most of these I devised prompts that invited literary analysis. When we read "Walking," for example, I asked students to take note of any boundaries, thresholds, or limits in the essay (whether material or metaphorical) and then to think about how transgressing those constraints would animate Thoreau's notion of wildness. This had special resonance with the fall 2020 class when they attended a virtual event with the author and activist Wen Stephenson, who drew on "Walking" and *Walden* to connect our fight for environmental justice with the nineteenth-century fight for abolition.

For the cluster on waterways, course material, again from a variety of media, included data on sea-level rise and flooding in the Midwest, news briefs on desiccated aquifers and rivers, and Paolo Bacigalupi's dystopian story "The Tamarisk Hunter." A focus on the Tennessee Valley Authority drew on Mississippi meander maps such as Harold Norman Fisk's 1944 rendering, archival photographs found on the TVA's own website, and Elia Kazan's *Wild River*. The film proved particularly generative, offering historical perspective on dam removals happening now and also eliciting conversation on topics as varied as the association of electrification with prosperity and how different domains of knowledge collide. Response options on the discussion boards included considering the effect of the documentary footage that opens the film, examining how wildness is projected onto "empty" spaces or onto women's bodies, elaborating the ecological implications of the matriarch's individualism, and discussing the oppression of Black citizens in the pre-electrified South.

Our third cluster brought together material on ice and glaciers, beginning with the advent of glaciology during the Romantic era and moving on to passages from Mary Shelley's *Frankenstein* (specifically, Walton's first three letters in vol. 1 and scenes of the Alpine sublime in vol. 2, ch. 1). I assigned portions of Wordsworth's "Essays upon Epitaphs" to help us respond to funeral services that have recently been held for vanishing glaciers. This inevitably took us to extinction and memorializing more broadly construed. Our in-class conversations explored how memorials confound the distinction between private and public (or between the individual and the aggregate) and what happens to voice when text becomes object. Considering ephemerality also allowed us to grapple with the complexities of scale inherent to our climate crisis. (Feedback loops seem to make time accelerate, but glaciers reify a time so "deep" as to seem halted.) In this cluster, students participated in two online discussions, one asking them to apply Wordsworth's ideas to a memorial that was meaningful to them, and the other asking them to take a position on heroism in the film for this cluster, Benedikt Erlingsson's *Woman at War*: Does the film, which engages with themes of ecofeminism, condone ecoterrorism? How does the conventional action script get flipped?

As this brief overview suggests, each cluster began with a problem (a common starting point at a polytechnic) and then elaborated its historical, technical, and cultural contexts through an eclectic grouping of materials—not only poems, short stories, personal and theoretical essays, documentaries, and narrative film but also infographics, scientific papers, radio spots, and archival documents. This capacious approach allowed the students to encounter literature from familiar entry points. (Electrifying the national grid, for example, will no doubt factor in the professional lives of several.) Our interpretive work with literary texts revealed key concepts, metanarratives, and habits of mind that continue to shape twenty-first-century responses to climate change, such as national identity; race, class, and gender; displacement and migration; and notions of justice.

Assignments

In addition to our discussion boards, three larger assignments sought points of intersection between the humanities and other fields. In fact, the first of these asked students to apply ecocritical principles to the output of their chosen professions. Deploying Peter Barry's description of what ecocritics do, their essays engaged with a series of questions about how their chosen fields take into account energy, mutuality, and collective responsibility for the environment and represent the natural world (Barry 270–71).

That last issue—representation—proved especially worthwhile. For example, after reviewing the principles of biophilic interior design, one student's essay generated useful questions for class discussion: What are the motivations for visually erasing the threshold between interior and exterior spaces? To what extent do representations of nature inside the home (on textiles and finishes, for example) promote biophilia? Might they instead aestheticize human domination? Another student, majoring in computer science, learned about how the Zanzibar Mapping Initiative has documented the local effects of climate change (see opendri.org/project/zanzibar/). This student proposed that in addition to fulfilling technical goals such as increasing efficiency to drive down energy usage, computer scientists can have additional impact as agents representing the natural world. In this admittedly experimental assignment, the object of ecocritical study was not literature, and the students' fields of study were not equally concerned with representations of the natural world. Yet, as these two student examples suggest, crossing disciplinary boundaries to apply ecocritical principles unconventionally can enhance self-awareness and clarify values. The assignment elicited innovative responses that reached beyond a more straightforward assessment of how a given industry engages with sustainability.

Our second long assignment was the book club project, a team effort. I grouped the students into four clubs, each of which read and presented on a single novel. In 2019 the texts were Bacigalupi's *The Water Knife*, Louise Erdrich's *Future Home of the Living God*, Barbara Kingsolver's *Flight Behavior*, and Ian McEwan's *Solar*. The next year, I substituted Lydia Millet's novel *A Children's Bible* for *The Water Knife*. We had already enjoyed a short story by Bacigalupi, and some of the *Water Knife* club found its violence problematic. What we gained with Millet's work was an opportunity to talk about massive systemic and generational failure.

The clubs met regularly outside class, and I asked them to add a social dimension whenever possible. I modeled this in 2019 by moving us to a much larger classroom one day so the clubs could enjoy more space, and I provided refreshments. One goal was to test whether social gatherings to discuss a climate change novel made it easier for the students to discuss climate change itself. The short answer is yes. Meeting in small groups outside the classroom apparently made it easier for students to share their opinions about, for example, the appetites of McEwan's main character, which then facilitated recognition of our own culture's gluttony. But most of the clubs reported having had only minimal conversational barriers to overcome. Any differences among club members seemed to have been interpretive, regarding, for instance, the role of religious faith in a work or the efficacy of a given author's approach (satire, realism, or postapocalyptic prophecy).

Opening with their texts' central climate-related concern, the clubs' presentations aimed to situate their novels in a multidisciplinary context. They considered current controversies, science, technology, and decision-making related to the book's primary concern. They then offered more explicitly literary analysis of character, theme, and tone; representation of human and nonhuman nature; relevant myths, traditions, or legacies; and, finally, whether and how empathy is elicited. In 2020, on *Zoom*, one club, clearly energized by their newly formed friendships, staged their presentation as a book review talk show.

End-of-semester surveys indicate that the book clubs were a highlight. Moreover, they were efficient; very few said they otherwise would have finished more than one book in a semester. Students appreciated the novels' capacity to convey fully realized futures in a climate-changed world. Even those works that are not obviously prophetic were assumed to offer a set of possible outcomes for our own lives—some more likely or terrifying than others. One conclusion, then, is that the book club assignment tested the usefulness of apocalyptic narrative. In contrast to *The Water Knife* and *Future Home of the Living God* (cautionary tales about a dangerously frayed civic structure and an assault on reproductive rights that could well characterize apocalypse), *Flight Behavior* and *A Children's Bible* seemed to meet an unspoken standard for plausibility that could inspire individual action in the reader. Those who read Kingsolver and Millet seemed to appreciate the pragmatism on display, a finding (generalized from a small sample set) that might redress some of the generic shortcomings Ghosh and others have identified.

The third longer assignment was an individual research project completed in two stages—a visual presentation of the work in progress and then, during exams, a written expansion upon that presentation. Each student chose a topic from broad categories I provided, these being (in 2020) heat, extractive industries, deforestation, and water in any form. Again they were asked to explore current controversies, pertinent science and technology, imminent decision-making, and (a new component) points of intersection with social justice. They also explored the cultural presence of their topic and created an annotated cultural bibliography of four works (literature, film, visual art) and four secondary sources.

The first time I taught the class, animals were also offered as a category, but a certain sameness characterized these projects, no doubt owing to the limited theoretical preparation I had provided. That said, one endangered species, the Sumatran tiger, did work successfully as a single element within a larger project investigating that island's deforestation. Here the student moved deftly from the threat of extinction and the crisis of biodiversity to orientalist representations in Western painting. His bibliography suggested that exoticizing nonhuman animal species prefigured our dominant culture's tendency to relegate climate effects to distant locales. A second successful example explored the extraction of stone, yielding an especially thought-provoking annotated bibliography that ranged from an autobiographical poem by James Dickey ("In the Marble Quarry") to charcoal drawings of stone carriers in Nazi concentration camps. The local setting of Dickey's poem offered a moving and stark contrast to the total alienation of the slave laborers. The aesthetic choices of these creators insist that we reckon with the aesthetic choices we make in design.

In this research project, and throughout the semester, students were not content simply to view literature as a springboard to more practical knowledge about the processes of the natural world. They sought to understand the provenance of vexing ideas that constrain our response to global warming, and they experienced the various ways literary texts help us interrogate our values. Displacement, environmental justice, biodiversity: challenges such as these became compelling—and, for some students, entirely new—considerations through their engagement with the literature of climate change. One engineering student arrived at this point quite eloquently in his final response paper, which asked students to articulate their guiding principles regarding global warming: it will take cultural work to recalibrate values, so the imagination must be fully empowered. That last prompt generated some of the best writing of the semester and helped to create coherence across the syllabus.

Especially in the fall of 2020, when the pandemic intensified for all of us a sense of multiple vulnerabilities, student feedback attested to the effectiveness of combining materials from many disciplines. In a survey with an eighty percent response rate, a bit more than a third of the respondents said they would not have taken the class if it had been a traditional literature class (a statistic that has many determinants). Highest marks were earned on survey questions related to adopting varying perspectives and honing interpretive skills. On the rather speculative question of whether

Wentworth should require a climate change course for graduation, three-quarters agreed. (The remaining quarter believed it should be a suggestion rather than a requirement.) As I continue to improve the syllabus of Reading the Weather, my hope is that multidisciplinary approaches will gain even greater traction—and will offer a powerful complement to discipline-specific literature classes. On my own campus, as we anticipate climate change and prepare for New Majority students (that is, the cohort of first-generation students, older adults, veterans, and students of color who now make up the majority of those entering college), the synergy of liberal learning and practical learning is more desirable than ever (Maimon).

Works Cited

Bacigalupi, Paolo. "The Tamarisk Hunter." *High Country News*, 26 June 2006, www.hcn.org/issues/325/tamarisk-hunter-Bacigalupi.

———. *The Water Knife*. Alfred A. Knopf, 2015.

Barry, Peter. *Beginning Theory: An Introduction to Literary and Cultural Theory*. 4th ed., Manchester UP, 2017.

Bewell, Alan. "Jefferson's Thermometer: Colonial Biogeographical Constructions of the Climate of America." *Romantic Science: The Literary Forms of Natural History*, edited by Noah Heringman, State U of New York P, 2003, pp. 111–38.

Cronon, William. "The Trouble with Wilderness; or, Getting Back to the Wrong Nature." *Environmental History*, vol. 1, no. 1, Jan. 1996, pp. 7–28. *JSTOR*, www.jstor.org/stable/3985059.

Crutzen, Paul J., and Eugene F. Stoermer. "The 'Anthropocene.'" *Global Change Newsletter*, no. 41, May 2000, pp. 17–18, www.igbp.net/download/18.316f18321323470177580001401/1376383088452/NL41.pdf.

Dickey, James L. "In the Marble Quarry." 1964. *Poetry Foundation*, www.poetryfoundation.org/poems/42721/in-the-marble-quarry.

Dowie, Mark. "The Myth of a Wilderness without Humans." *The MIT Press Reader*, 11 Oct. 2019, thereader.mitpress.mit.edu/the-myth-of-a-wilderness-without-humans/.

Erdrich, Louise. *Future Home of the Living God*. HarperCollins Publishers, 2017.

Erlingsson, Benedikt. *Woman at War*. Magnolia, 2018.

Fisk, Harold Norman. "Ancient Courses [of the] Mississippi Meander Belt, Sheet 9." 1944. *The History of America in One Hundred Maps*, by Susan Schulten, U of Chicago P, 2018, p. 218.

Hester, Jessica Leigh. "Where Are Europe's Last Fairytale Forests?" *Atlas Obscura*, 1 June 2018, www.atlasobscura.com/articles/mapping-europes-primeval-forests.

Howe, Joshua P. "This Is Nature; This Is Un-Nature: Reading the Keeling Curve." *Environmental History*, vol. 20, no. 22, 2015, pp. 286–93, https://doi.org/10.1093/envhis/emv005.

Kaplan, Jed O., et al. "The Prehistoric and Preindustrial Deforestation of Europe." *Quarternary Science Review*, no. 28, 2009, pp. 3016–34. *Elsevier*, www.wsl.ch/staff/niklaus.zimmermann/papers/QuatSciRev_Kaplan_2009.pdf.

Kazan, Elia. *Wild River.* Twentieth Century–Fox, 1960.

Kingsolver, Barbara. *Flight Behavior.* HarperCollins Publishers, 2012.

Lewis, Simon L., and Mark A. Maslin. "Defining the Anthropocene." *Nature*, vol. 519, 12 Mar. 2015, pp. 171–80, https://doi.org/10/1038/nature14258.

Maimon, Elaine. "One Year of COVID: Higher Ed Will Never Be the Same Again." *The Philadelphia Citizen*, 22 Mar. 2021, thephiladelphiacitizen.org/higher-ed-never-same/.

Marinelli, Janet. "In the Sierras, New Approaches to Protecting Forests under Stress." *Yale Environment 360*, 13 Feb. 2017, e360.yale.edu/features/in-the-sierras-new-thinking-on-protecting-forests-under-stress.

McEwan, Ian. *Solar.* Random House, 2010.

McLane, Maureen N. "Märchen." *The Kenyon Review*, new series, vol. 33, no. 3, summer 2011, p. 41. *JSTOR*, www.jstor.org/stable/41304640.

———. "OK Let's Go." *Poets.org*, 5 July 2016, poets.org/poem/ok-lets-go.

Millet, Lydia. *A Children's Bible.* W. W. Norton, 2020.

Rittel, Horst W. J., and Melvin M. Webber. "Dilemmas in a General Theory of Planning." *Policy Sciences*, vol. 4, 1973, pp. 155–69. *Springer Link*, https://doi.org/10.1007/BF01405730.

Robbins, Jim. "As Mass Timber Takes Off, How Green Is This New Building Material?" *Yale Environment 360*, 9 Apr. 2019, e360.yale.edu/features/as-mass-timber-takes-off-how-green-is-this-new-building-material.

Shelley, Mary. *Frankenstein.* 1818. Edited by Stuart Curran, *Romantic Circles*, 2009, romantic-circles.org/editions/frankenstein/1818_contents.html.

Snyder, Gary. "Mid-August at Sourdough Mountain Lookout." 1992. *Poetry Foundation*, www.poetryfoundation.org/poems/47182/mid-august-at-sourdough-mountain-lookout.

Thoreau, Henry David. "Walking." *The Atlantic Monthly*, June 1862, theatlantic.com/magazine/archive/1862/06/walking/304674/.

Whyte, Kyle Powys. "White Allies, Let's Be Honest about Decolonization." *Yes!*, 3 Apr. 2018, yesmagazine.org/issue/decolonize/2018/04/03/white-allies-lets-be-honest-about-decolonization/.

Wordsworth, William. "Essays upon Epitaphs." *The Prose Works of William Wordsworth*, vol. 2, edited by W. J. B. Owen and Jane Worthington Smyser, Oxford UP, 1974, pp. 24–76. *Oxford Scholarly Editions Online*, www.oxfordscholarlyeditions.com/view/10.1093/actrade/9780198719755.book.1/actrade-9780198719755-work-2.

Ali Brox

Imagining Just Futures:
Teaching the Literature of
Climate Change as Social Responsibility

"I can't believe I thought that!" a student in the back of the room ex-
claimed. It was the last day of my first-year seminar about climate change
literature. I had just returned an initial assessment questionnaire that students had taken on the first day of class and asked if there were any answers the students would change now, at the end of the course. The student was reacting to her response of "not very important" to the question
"How important do you think race is when considering climate change?"
Leaving the classroom that day, I felt that at least one student had achieved
a major learning goal. As Paolo Bacigalupi writes,

> Sometimes, I've discovered, it's possible for a fiction writer to perform
> a kind of hack on a reader's mind, making them feel things that do
> not yet exist. And if we writers do our jobs well, when the reader closes
> a book, they will see the world differently. They will see low water lev-
> els in Lake Mead and connect them to catastrophic loss, to forced
> migration, to uncertainty. (xiv)

My student saw the world differently at the end of the course.

In my environmental studies courses, I approach the teaching of climate
change literature through the lenses of hope and despair, with a focus on

texts by women of color, such as Octavia E. Butler's *Parable of the Sower*, N. K. Jemisin's *The Fifth Season*, and Sherri L. Smith's *Orleans*. Studies have shown that fiction readers display more empathy to other people and places (Oatley). The development of empathy is crucial in preparing students to be ethical global citizens, and climate change literature can be a particularly effective mode for exploring diverse perspectives and empathetic scenarios. In *Climate Change Fictions*, Antonia Mehnert explains, "[C]ultural texts pose ethical questions concerning not only the extent of human impact on the planet but also decisions about who or what is saved or left to die in a climatically changed future. They therefore engage in a 'poetics of responsibility,' that is, a discussion about the responsibility humankind has towards its own actions" (2–3). By emphasizing environmental justice issues and the disproportionate effects of climate change on certain people and places, instructors can help students grasp the international challenges posed by a warming planet—particularly questions of what types of futures are imagined and for whom. Teaching the literature of climate change this way, I have witnessed developments in students' understanding of the social, political, and cultural implications of global warming.

This essay provides strategies and practical examples for teaching the literature of climate change in two types of courses. The first model, a first-year seminar, focuses on climate change literature during a semester-length course. This course examines novels, films, short stories, and scholarly articles. Students gain critical thinking skills through a series of assignments that lead to a final research project. The second model, an introductory environmental studies course required for majors, integrates the literature of climate change into a unit that complements other environmental topics and issues. The unit involves a novel, background readings about literary criticism, and a visit to the campus research library. These materials introduce students to the ways the humanities shape our conception of global warming and prompt students to reflect on the social conflicts that are heightened in climate-changed futures. The approaches I describe could be extrapolated to other classes dedicated to literature or environmental themes, depending on the needs of instructors and students. Regardless of the amount of course time one can devote to the literature of climate change, exposing students to environmental humanities perspectives proves useful for achieving broader learning outcomes that we as a faculty hope university graduates possess.

The literature of climate change is an ideal medium for reflecting on the history of humans' relationships to their environments, speculating on possible alternatives to current practices, and imagining futures of restoration or destruction. It offers opportunities to develop students' appreciation for cultural and global diversity and to cultivate their ethical integrity and social responsibility. These learning outcomes are most effectively realized when students are exposed to the study of the literary imagination:

> A narrative . . . is different from a scientific account in that it will allow those who receive it to imagine that sense or feeling of what it is like to lose one's footing on slippery ice. It will cue recipients—be they viewers, listeners, or readers—to simulate that sense or feeling in their minds, using their own real-world experiences as models and their own bodies as sounding boards for the simulation. (Weik von Mossner 7)

This simulation is crucial for imagining what others experience and can lead to increased empathy and a sense of responsibility to take action.

Literary critics have pointed out the importance of the humanities in debates about climate change, an issue that too often is seen as the sole purview of the natural sciences. John Joseph Adams, in his introduction to *Loosed upon the World: The Saga Anthology of Climate Fiction*, explains the role literature can play: "Fiction is a powerful tool for helping us contextualize the world around us. By approaching the topic [climate change] in the realm of fiction, we can perhaps humanize and illuminate the issue in ways that aren't as easy to do with only science and cold equations" (xii). Steven Allison and Tyrus Miller agree with Adams about how the humanities can touch people in ways science may not: "So far, scientific facts have not motivated Americans to support the huge societal transformations needed to stop climate change. . . . By tapping into what moves people, the emerging field of environmental humanities can help spur climate action." In addition to exciting or inspiring people, literature raises important questions about what futures are possible and desired. Mehnert argues, incorporating a quotation from the geography professor Mike Hulme, that literature is a necessary component for considering the ethical implications of climate change: "[A]n awareness of the social and cultural narratives, which shape the very idea and interpretations of climate change, is essential to grasp the full complexity of climate change, and, in particular, 'to attend more closely to what we really want to achieve for humanity'" (8). That is where the power of the literary imagination lies; if we want a more just and equitable future with clean air, clean water, and sustainable energy and food, then we must imagine it first. Crucial to that process is rethinking current policy and practice.

In my course for environmental studies majors, I have found that in order to overcome students' resistance to reading a novel it is necessary to make the connection between literature and attitudes about the environment and nature. Once students are exposed to a brief history of the role literature has played in environmental thought, they begin to understand how values, beliefs, and actions are shaped by discourse and how environmental discourses are culturally specific and change over time. Kaci Zarek, a student in my introductory environmental studies course during the spring 2019 semester, made the connection between aspects of the novel *Orleans* and current events:

> I think this book shows the harsh realities of what climate change can do in the coming years. We have already experienced hurricane Katrina and [seen] its devastating effects in Louisiana. This book shows how fundamental it is for more representatives to create proposals like the Green New Deal, which brings to light the environmental disasters that will happen. Also, *Orleans* demonstrates the environmental injustice between the government and vulnerable communities. For example, how the government is resolving Flint, Michigan's water problem.

Zarek associates issues raised by *Orleans* with specific government policies. The future world in the novel serves as a way for Zarek to reflect on existing practices and advocate for actions to mitigate "harsh realities" or prevent them from coming to fruition. Bacigalupi explains how writers achieve this: "It's interesting that by creating a made-up world, you can show the *real* world more sharply and clearly, and in that process, you have the chance of making people engage not with the future, but with the intense realities of our present—the realities that were previously passing them by" (xiv). For some of my students, one of those realities is that climate change affects people differently based on geographic location, gender, race, ethnicity, and economic class. Many do not recognize the effects of climate change in their daily lives, and thus climate change is still geographically or temporally distant. However, we know climate change is happening now. I teach climate fiction to expose students to perspectives of those disproportionately affected by the consequences of climate change in the present and projected future. The literature I assign represents women, people of color, the poor, and immigrants in their imagined worlds.

While I encountered more students who were familiar with cli-fi in 2021 than I did when I first started teaching this material in 2018, it continues to be necessary to introduce definitions of climate change literature and common characteristics of the genre. I do this by assigning a *YouTube* video in which Stephanie LeMenager provides the following definition:

> Cli-fi means climate fiction. It's actually a vast array of different kinds of literary products. Everything from memoirs . . . to novels to youth fictions . . . to film. So climate fiction is a very broad multimedia, multigenre project that many artists and filmmakers and authors have been pursuing in an attempt to tell the story of climate change. ("Stephanie LeMenager")

Once we have a baseline for what climate change literature is and what its basic components are, we begin discussions about examples students already may be familiar with, such as movies like *Wall-E* and *Snowpiercer*. Then we begin applying the definitions and characteristics to course readings. The first time I taught a course dedicated solely to climate change literature, I started by assigning reading from the scholarly conversation around cli-fi in order to expose students to what others say about the literature as well as to model part of the research process. While I still integrate scholarly articles in my instruction, now I start with short stories or memoir essays: the theoretical readings were too much at the beginning and were not the elements that excited students about the genre. Currently, students read short, primary text examples like "Outer Rims," by Toiya Kristen Finley; Margaret Atwood's "Time Capsule Found on the Dead Planet"; and Zadie Smith's "Elegy for a Country's Seasons." These have proven effective in familiarizing students with some of the main themes and characteristics of cli-fi before they tackle the scholarly articles and longer primary texts.

One of my main goals when teaching climate change literature is to expose students to perspectives of those disproportionally affected by the consequences of climate change. When teaching Butler's *Parable of the Sower*, I provide questions to focus our reading and discussions on achieving that goal:

> What futures are imagined in cli-fi for those who suffer the most from the burdens of climate change?
> How do humans persist in the climate-changed worlds that are imagined? How would a different future materialize?
> Why is cli-fi important? What can it convey that other forms of knowledge (such as science and data) cannot?

To demonstrate for students ways to answer the first question, I lecture about environmental justice, using a case study about Hurricane Katrina.

Then I model analysis of passages from *Parable of the Sower* that illustrate comparable viewpoints. For example,

> There's a big, early season storm blowing itself out in the Gulf of Mexico. It's bounced around the Gulf, killing people from Florida to Texas and down into Mexico. There are over 700 known dead so far. One hurricane. And how many people has it hurt? How many are going to starve later because of destroyed crops? That's nature. Is it God? Most of the dead are the street poor who have nowhere to go and who don't hear the warnings until it's too late for their feet to take them to safety. Where's safety for them anyway? Is it a sin against God to be poor? We're almost poor ourselves. There are fewer and fewer jobs among us, more of us being born, more kids growing up with nothing to look forward to. One way or another, we'll all be poor some day. (Butler 15)

After I explain how this exemplifies an environmental justice perspective, I ask students to find two additional examples from the novel that represent similar perspectives. Students cite the passages or scenes and explain how they fit within an environmental justice framework.

I pair *Parable of the Sower* with the "Hyperempathy" entry by Rebecca Evans from *An Ecotopian Lexicon*. Lauren Olamina, the protagonist in Butler's novel, suffers from the hyperempathy condition because her mother used a drug while pregnant. A "sharer," Lauren physically experiences the pain or pleasure that she witnesses in others. In her entry, Evans explores the possibilities hyperempathy represents: "Sharers embody not sympathy but radical empathy, which prompts moral action and avoids a reliance on a distant, speculative, and often pitying relation to those who suffer. Hyperempathy thus exemplifies an emotional orientation that may be necessary to catalyze action in the face of environmental and climate injustice" (117). Alexa Weik von Mossner examines the simulated experience of reading and how it corresponds to empathy: "[T]here is a political value to embodied simulation since cognitive and affective empathy—and the resulting imaginary experience of another's situation—are the very basis for recognition and a resulting desire for the improvement of that situation" (171). Both Evans and Weik von Mossner focus on the actions that can result from empathy. To date, I have not followed up on the behaviors my students may enact or change after reading the materials in my courses, but Matthew Schneider-Mayerson has researched the influence of climate fiction through an empirical survey of readers. His results suggest that many readers drew connections between reading climate fiction and subsequent actions (493). These actions tended to fall into two camps: conversations with friends or family about climate change and changes in individual behavior, like using recyclable shopping bags. While "many of these new behaviors barely contribute to climate mitigation, collective adaptation, or the pursuit of climate justice" (494), the influence of climate literature on readers' awareness and actions should not be dismissed and deserves further study.

Student work indicates that the environmental justice lens through which my courses approach climate change literature has successfully cultivated awareness of and empathy for environmental injustices. In her final presentation for my fall 2020 first-year seminar course, Liz Collins explained that she had learned how "cli-fi can be used to bring up other issues like environmental racism and how that is portrayed in *Parable of the Sower* and how we see it in everyday life as well." Collins's comment reflects her recognition that environmental injustices extend beyond the pages of the texts assigned in class. Skyler Clark echoes this with a similar realization about inequities in her final paper for the fall 2018 version of the same course:

> It is also worth noting that a majority of the characters are people of color, and they seem to be worse off compared to their white counterparts. Minority characters have been enslaved, raped, abandoned, and sold into prostitution, and many of them are impoverished, which incorporates the social themes seen in cli-fi, as it is understood that climate change will most directly affect those who live in poverty.

Clark's reference to the social themes in cli-fi illustrates her understanding that the novel emphasizes how the effects of climate change extend beyond the physical environment and serve to heighten social concerns like racism and poverty. Collins and Clark are representative of students who have developed greater understanding of the ethical implications of global warming through engagement with climate change literature.

There are many reasons to teach the literature of climate change, not least of which is the growing list of exciting texts. While the content alone might motivate instructors and students alike, as an educator, I am constantly considering what responsibilities I have to my students in the midst of human-caused climate change. What do I teach? What skills must I prepare students with for the future? What can protect them from the dire predictions scientists and some policymakers espouse, or what can enable their resilience? These questions, and others, reinforce my commitment to teaching climate change literature. I am fortunate to have environmental studies students who are dedicated to comprehending and solving some of our planet's biggest challenges. The abilities they develop by critically examining the literature of climate change through an environmental justice lens serve as an additional tool for imagining and manifesting the best future possible for everyone and every being.

Works Cited

Adams, John Joseph. Introduction. Adams, pp. xi–xii.

———. editor. *Loosed upon the World: The Saga Anthology of Climate Fiction.* Saga Press, 2015.

Allison, Steven D., and Tyrus Miller. "Why Science Needs the Humanities to Solve Climate Change." *The Conversation*, 1 Aug. 2019, theconversation .com/why-science-needs-the-humanities-to-solve-climate-change-113832.

Atwood, Margaret. "Time Capsule Found on the Dead Planet." Adams, pp. 556–58.

Bacigalupi, Paolo. Foreword. Adams, pp. xiii–xvii.

Butler, Octavia E. *Parable of the Sower.* Grand Central Publishing, 2000.

Evans, Rebecca. "Hyperempathy." *An Ecotopian Lexicon*, edited by Matthew Schneider-Mayerson and Brent Ryan Bellamy, U of Minnesota P, 2019.

Finley, Toiya Kristen. "Outer Rims." Adams, pp. 39–51.

Jemisin, N. K. *The Fifth Season.* Orbit, 2015.

Mehnert, Antonia. *Climate Change Fictions: Representations of Global Warming in American Literature.* Palgrave Macmillan, 2016.

Oatley, Keith. "Fiction: Simulation of Social Worlds." *Trends in Cognitive Sciences*, vol. 20, no. 8, 2016, pp. 618–28. *Science Direct*, https://doi.org/10 .1016/j.tics.2016.06.002.

Schneider-Mayerson, Matthew. "The Influence of Climate Fiction: An Empirical Survey of Readers." *Environmental Humanities*, vol. 10, no. 2, Nov. 2018, pp. 473–500, https://doi.org/10.1215/22011919-7156848.

Smith, Zadie. "Elegy for a Country's Seasons." *The New York Review of Books*, 3 Apr. 2014, www.nybooks.com/articles/2014/04/03/elegy-countrys -seasons/.

"Stephanie LeMenager: What Is Cli-Fi? Radcliffe Institute." *YouTube*, uploaded by Harvard University, 25 Jan. 2017, www.youtube.com/watch?v =P9XuxHtfOxQ.

Weik von Mossner, Alexa. *Affective Ecologies: Empathy, Emotion, and Environmental Narrative.* Ohio State UP, 2017.

Debra J. Rosenthal and Jeffrey Johansen

Cli-Fi Linked to a Climate Science Course

This essay describes an innovative interdisciplinary way to teach an introductory-level literature course on climate change fiction by linking it back-to-back to a biology class on the science of anthropogenic climate change. We are a professor of literature (Rosenthal) and a professor of biology (Johansen); we have taught this linked pair of courses every semester since 2017. We believe that our students benefit tremendously from the opportunity to learn basic climate science and then see how concerns about planetary catastrophe can be effectively expressed by writers.

At John Carroll University in Cleveland, Ohio, students must fulfill part of their core graduation requirements by taking a linked pair of classes (from two different departments) that professors design around a common theme. We have designed such a linked pair in which students must co-register for both Rosenthal's English class and Johansen's biology class; students thus engage in six hours per week of instruction about climate change. In the biology class, students learn about carbon dioxide emissions, the ice albedo effect, the hockey stick curve, rising sea levels, ocean acidification, desertification, and more. In the English class, students learn that stories hold a particular power: storytelling lets us inhabit and identify with the reality and truth that climate science predicts and explains.

220

The biology course in our linked pair, designed for students not majoring in science, focuses on theories and predictions of climate science and evidence that the climate has started a period of rapid and deleterious change. We hope that by the end of the course students will be able to explain the theory behind anthropogenic climate change, relate the primary evidence that such change has already occurred, articulate the reasons most climate scientists conclude that the majority of change detected so far is mostly anthropogenic, and interpret predictions for future change, particularly those changes forecast for North America.

The diverse region-specific impacts of climate change can best be understood by developing a familiarity of place so that the predicted devastation can be more deeply felt. Since it is hard for most Americans, and indeed for most of the world, to understand environmental change and what it will mean in diverse regions of the world, literature can help us feel what it is like to live in and inhabit a specific place. Since we cannot begin to comprehend the catastrophic loss of the Amazon rainforest or the loss of the Great Barrier Reef if we have not experienced either place, works of literature can help transport us to those locales. Likewise, many of us cannot understand predicted changes even in our own bioregion because our urban and suburban environments insulate us from interacting closely with the natural world.

We have organized our syllabi according to bioregion so that our courses generally parallel each other as students learn how climate change will affect specific areas of the United States. For example, while Johansen teaches about climate change in the Northeast and New England, Rosenthal's students read *Walden*, by Henry David Thoreau. Some very interesting work has been done recently by Richard Primack in his book *Walden Warming*; by using Thoreau's careful counting of butterflies, birds, and flowering plants as a data set, environmentalists today can measure the changing climate around Walden Pond. Thoreau thus becomes almost a coauthor in scientific studies of flora and fauna in twenty-first-century Massachusetts. Reading *Walden* helps students understand species' evolutionary history, and Thoreau's journals and essays help them see that extinction risk is directly correlated with traits that are known to be affected by changes in climate. Students can more deeply resonate with the pertinent biology question "What happens to ecosystems when species are extirpated or go extinct?" after discussing the mystical transcendental experiences Thoreau enjoyed next to his beloved pond.

The next bioregion studied is the Midwest. Johansen teaches students about the importance of agriculture to the Midwest, noting that, while increased global temperatures will beneficially prolong the growing season, the increasing severity of weather events will lead to a net decrease in agricultural productivity. Climate change will exacerbate a range of risks to the Great Lakes, including the range and distribution of certain fish species, increased invasive species, harmful algal blooms, and declining beach health. Students become deeply concerned when they learn that climate models predict increased heat stress, flooding, and drought in the Midwest and that these effects will be multiplied by new crop pests and disease, increased competition from nonnative species, atmospheric pollutants, and, most concerning, economic shocks such as crop failures or reduced yields due to extreme weather events.

While studying these environmental predictions, students read in Rosenthal's class Aldo Leopold's *A Sand County Almanac* and a short story by Ernest Hemingway, usually "Big Two-Hearted River." These two works' deep descriptions of Wisconsin and Upper Michigan help students picture the bioregion. We study Leopold's "land ethic" and his advocacy for stewardship and a mutually sustaining relationship between people and their environment. Rosenthal jokingly warns students before reading "Big Two-Hearted River" that nothing happens plotwise other than "man goes fishing." Yet students dive into the story, absorb Nick Adams's respect for the river, and learn how immersion in the river can heal. The healthy ecosystem helps Nick adapt after returning from the trauma of World War I just as the grasshoppers must adapt after a scorching fire.

After studying the Midwest, we direct our focus to the Southeast. Literature by southern writers helps students imagine what is at stake as global temperatures rise and affect their home turf. To help students inhabit the South, Rosenthal's class usually spends three to four weeks on Barbara Kingsolver's *Flight Behavior*. The plot revolves around a working-class family on whose property several million monarch butterflies wrongly overwinter because climate heating has confused their internal map as they migrate from Canada to Mexico. Kingsolver invites readers into the lives of the working poor in Appalachia as they grapple with climate change, economic decisions, and the migratory error of the monarchs. In the paired biology course, students study how the Southeast will be affected by heat waves that can kill the elderly and the poor, crop failure (especially when high temperatures combine with drought), increased fire intensity and frequency, reduction in dairy and livestock production, and coastal erosion caused by rising sea levels combined with storm surges and increased hurricane activity.

Finally, our syllabi turn toward the Southwest. Rosenthal teaches Paolo Bacigalupi's short story "The Tamarisk Hunter" and Olivia Clare's "Eye of Water." Both address the desertification of the Southwest, where water is rigidly allocated. These stories help students imagine the region as they learn about the predicted effects of increased heat, drought, and wildfires on crops and humans.

We assign two projects that require students to draw connections between the disciplines of biology and English: a paper on how climate change will affect their future careers and a movie-pitch final project.

To encourage our students to reflect on the likely impact of climate change on their future, and then to manage their resulting sober and conflicted emotions, we have created an assignment we call "How Will Climate Change Impact My Career?" The assignment first asks students to define their career area. If students are not yet sure about their career aspirations, we ask that they identify some aspect of a vocation that they gravitate toward (for example, a helping profession, a STEM-related field, or environmental work of some kind). If students express interest in an emerging field, we ask them to describe briefly what a professional in that career might do.

Once students have defined or described their career area, we ask them to cast forward to life after graduation. Using the science of climate change they have studied all semester, and the ways the literary imagination has helped them think of a threatened human existence, students must predict how global heating will affect their chosen careers. Their papers usually express what is known as ecological grief or ecoanxiety—feelings of despair and anxiety that come with realizing the future personal impact of our fossil fuel culture. After students turn in their papers, we hold an in-class discussion to help them reflect on their emotional responses to ecological degradation, social injustice, and their vocational choices. We discuss students' anxious feelings about their career-choice papers in relation to the emotional registers that fiction writers and poets elicit in order to raise readers' awareness of planetary catastrophe and motivate them to be climate advocates. We try to see how literature can imbue existential doubt with narrative purpose.

For our summative final linked project that draws upon both disciplines, students must make a *PowerPoint* presentation of a pitch for a Hollywood-style movie blockbuster about climate change. We pretend to be Hollywood movie executives; students engage in a friendly competition to see whose movie pitch will "win" millions of dollars in funding.

This assignment is challenging because students must propose plots for entertaining movies that would run at least ninety minutes and would present correct climate science. A pitch reminiscent of *Waterworld*, in which Earth's polar ice caps have melted and the oceans have submerged the continents, would not receive a good grade in our class, for example, because no Intergovernmental Panel on Climate Change report or scientific evidence could be cited to support such sea-level rise. Similarly, the film *Interstellar* would not receive a high grade, because there is no scientific evidence to prove that climate change would cause a fungus capable of wiping out all of the world's crops. To earn a good grade, student movie pitches must have a compelling plot and be scientifically sound.

Students present their movie pitches during the last two days of the semester. It is fun to end the course seeing the students creatively combining scientific fact with compelling story lines. Students vote anonymously for a winner and runners-up. The prize, aside from bragging rights, is a copy of Elizabeth Kolbert's *The Sixth Extinction: An Unnatural History*. Over the years students have pitched a wide range of genres: thrillers, disasters, animated children's movies, rom-coms, science fiction, and so on. In linking a STEM course to a literature course we have found that students learn the power of facts as well as the power of storytelling to communicate about climate change.

Works Cited

Bacigalupi, Paolo. "The Tamarisk Hunter." *Loosed upon the World: The Saga Anthology of Climate Fiction*, edited by John Joseph Adams, Saga Press, 2015, pp. 511–26.

Clare, Olivia. "Eye of Water." *Disasters in the First World*, by Clare, Grove Press, 2017, pp. 169–86.

Hemingway, Ernest. "Big Two-Hearted River." 1925. *In Our Time*, by Hemingway, edited by J. Gerald Kennedy, Norton Critical Edition, W. W. Norton, 2022, pp. 177–99.

Interstellar. Directed by Christopher Nolan, Paramount Pictures, 2014.

Kingsolver, Barbara. *Flight Behavior*. HarperCollins Publishers, 2012.

Kolbert, Elizabeth. *The Sixth Extinction: An Unnatural History*. Holt, 2014.

Leopold, Aldo. *A Sand County Almanac*. Ballantine, 1986.

Primack, Richard. *Walden Warming: Climate Change Comes to Thoreau's Woods*. U of Chicago P, 2015.

Thoreau, Henry David. *Walden; or, Life in the Woods*. 1854. *Lit2Go*, U of South Florida, etc.usf.edu/lit2go/90/walden-or-life-in-the-woods/.

Waterworld. Directed by Kevin Reynolds, Universal Pictures, 1995.

Ben Jamieson Stanley and Emily S. Davis

Climate Fiction and the Global South

Coined by the journalist Dan Bloom, *climate fiction*, or *cli-fi*, has become a prominent term in media (as tracked in Bloom's *Cli-Fi Report*) and in scholarship on British, American, and Canadian literatures. Fictions from the Global South are rarely centered in this cli-fi conversation, even though two of the thinkers commonly cited on the epistemic and narrative challenges of climate change—the historian Dipesh Chakrabarty and the novelist Amitav Ghosh—are canonical to postcolonial studies.[1] What might a focus on Global South literature offer to teaching climate fiction? And what might an emphasis on climate change offer to teaching literatures of migration?

We began thinking about these questions together when planning a pedagogical collaboration for fall 2020. As the crises of that year unfolded—the COVID-19 pandemic, detentions of migrants at the US-Mexico border, and a reckoning with anti-Black police violence in the United States—we recognized the urgency of addressing the sometimes separated issues of climate change and racial justice. What might it look like, we asked, to teach climate fiction in a way that centers Global South authors, as well as Black and Indigenous authors from the Global North, and combines the energy around the climate fiction category with the insights of postcolonial studies? The following essay reflects on this question from our perspectives as two white settler scholars trained in postcolonial studies, both teaching at the same primarily white institution in the United States.

A postcolonial approach means more than teaching a few Global South texts. We wanted to offer students frameworks for challenging US-centric narratives about migration and climate change. Centering Global South thinkers can recast climate change not just as a universal problem that affects "the whole planet" but also as an environmental justice issue that affects different nations and populations unevenly. This means attending to interplay with other injustices: how Pacific Islanders impacted by nuclear testing might experience sea-level rise as a new manifestation of environmental imperialism, for example, or how amplified storms might exacerbate precarity in Black communities in the American South. As these examples suggest, climate change is enmeshed with domestic and transnational racial injustice. The UN Paris Agreement holds all countries responsible for climate change, yet the Global North and multinational corporations arguably owe a "historic climate debt" to the Global South (Satgar and Cherry 319): the Global North has subsidized polluting industries and enabled wealthy people to enjoy refrigerators and private cars and would deny those comforts to those in the Global South. Not only have poorer countries caused fewer emissions, they have fewer funds for seawalls, levees, and relocation projects, whereas "wealthy nations will fund their adaptation . . . with capital gained from a century of exploiting fossil fuels" (Schneider-Mayerson 948). Postcolonial approaches can help students think globally about what these injustices should mean for mitigation and adaptation. This will require students—especially white American students—to engage with perspectives that challenge their own, even if contained under a shared umbrella of climate concern.

Our collaboration began with a plan to teach two coordinated classes: a course on migration literature taught by Emily and another on climate fiction taught by Ben, intersecting in their concerns with climate-related migration. Although COVID-19 obligated us to scale back plans, our classes read two shared texts—Cherie Dimaline's *The Marrow Thieves* and Amitav Ghosh's *The Hungry Tide*—and exchanged perspectives in groups online. In what follows, we share our takeaways from this experiment, offering suggestions for other educators about how Global South and postcolonial frameworks might infuse climate fiction teaching.

Climate Fiction and the Postcolonial Environmental Humanities (Ben)

As a scholar of postcolonial studies and environmental humanities, I tend to organize courses under rubrics such as "global environmental justice" or "environment and empire." But the urgency of climate change and the ascendancy of the "climate fiction" discourse have prompted me to explore teaching with this term. I also want to put pressure on it. Teaching cli-fi from the angle of postcolonial environmental humanities raises generative questions about how we should define climate fiction and about the rubric's affordances and limitations.

Scholars have warned against defining cli-fi as a coherent, singular genre (Evans 95). Yet cli-fi can feel differently narrow in its literary geographies. For example, a special issue of *Studies in the Novel* on "climate change fiction" addresses a range of genres—sci-fi, realism, graphic novel, romance—but, with notable exceptions, most contributor essays privilege American, British, and Canadian authors (Craps and Crownshaw). So do many studies of climate fiction or Anthropocene fiction (Johns-Putra; Trexler; Marshall; Schneider-Mayerson; LeMenager). Some do engage with colonialism and race: for example, Stephanie LeMenager begins a genealogy of cli-fi with African American and Native writers who foreground that climate change is continuous with "colonialist practice, including racism, the first act of enslavement . . . , the first rape, the first territorial theft" (227). LeMenager decenters white settlers and offers a conceptual framework applicable beyond the North, but she does not highlight Global South writers.

But although the climate fiction conversation has skewed northward, I hesitate to suggest that we simply "add" authors from the Global South. It is worth considering whether climate fiction as currently defined is genuinely a better descriptor for Global North texts. This is not to suggest that Global South fictions have nothing to say about climate change but rather that their approaches may elude a rubric developed with Global North texts in mind. Climate fiction is often associated with depictions of future disaster, and this temporality may be less relevant for writers from contexts where climate impacts are already severe. Likewise, that climate change affects "all humanity" might not be the priority for writers concerned with uneven climate risks.[2]

Perhaps in part for these reasons, many postcolonial ecocritics foreground rubrics other than climate fiction.[3] Global South novels might themselves suggest different frameworks. For example, Ghosh's *The Hungry Tide* does mention geological timescales, sea-level rise, and other themes associated with climate fiction, but what has made the novel canonical to postcolonial ecocriticism is its attention to how Western-style conservation projects steamroll subaltern needs and exacerbate refugee crises. Rather than suppose that *The Hungry Tide* fails to represent climate change—as Ghosh asserts in *The Great Derangement* (9)—I would suggest that the novel's climate relevance emerges through our recognition that climate change is continuous with histories of colonization, displacement, extraction, and exploitation, which benefit some peoples and places at the expense of others. Recognizing such connections is crucial for students exploring literature and climate change and can be encouraged through Global South texts.

To engage Global South writers, then, may put some interesting pressure on climate fiction and may entail cross-pollinating with other rubrics. In my course Climate Change and Environmental Refugees, I taught "climate fiction?" as an open question and introduced *environmental refugeeism* as a second term. Global North climate fictions such as Paolo Bacigalupi's *The Water Knife* predict (quite realistically) that climate change will displace numerous people in the future. However, climate change is displacing people already, especially in the Global South. Foregrounding both climate fiction and refugeeism allowed my students and me to think critically about how globalization, migration, and climate change all connect; how locating climate disruption in the future might be a privileged perspective; and how climate representation might differ in texts that foreground historical, present, and recurring modes of displacement and injustice (such as *The Hungry Tide*, *The Marrow Thieves*, and Kathy Jetñil-Kijiner's *Iep Jāltok*).

I also wanted to create space for students to wrestle with defining climate fiction themselves in culturally relevant ways. In an assignment called "Proposing a New Text," each student wrote a book or film review suggesting an addition to our syllabus. I provided a list of considerations, such as relevance to climate change and environmental refugeeism; what perspective, geography, and genre would be represented; practicality; and access. The class then voted on several texts to actually add to the syllabus; we read these together in the last two weeks of the course. This experiment was fruitful in part because the results were mixed. Some students vaguely mentioned environmental themes, leading me to wonder if our conversations had become too diffuse. But others interrogated either how their text foregrounded climate change and environmental refugee-ism or how and why its inclusion would put pressure on those terms and broaden our thinking. Exploring multiple key terms might have made my students less attuned to the specificity of climate change than if we had defined climate fiction more straightforwardly, yet I believe it was more important to tackle canonicity, bias, globality, and environmental justice than to nail down our definition of climate fiction.

I continue to wonder whether climate fiction is a useful term when teaching Global South literatures. If defined too loosely, cli-fi could become a synonym for the umbrella category of environmental literature, sacrificing the specificity of climate change. But codifying the term may privilege northern perspectives. Can we keep climate fiction as a specific and teachable concept yet continue to decolonize syllabi, amplify marginalized perspectives, and recenter climate justice? I do not have a definitive answer but found it tremendously generative to explore such questions with Emily and with our students.

Migration Literature, Climate Change, and the Global South (Emily)

The collaboration between my class and Ben's found its meeting point in the conceptual framework of the "environmental refugee." My work for the past few years has pursued an interdisciplinary approach to the study of literature and migration, drawing heavily from international human rights law, postcolonial studies, and globalization studies and actualized in the public-facing website *Moving Fictions: Exploring Migration in Modern Literature* (sites.udel.edu/movingfictions/). I had heard the term *climate fiction*, but it was not a central term in my analysis of the representation strategies and geopolitics of migration. While Ben sought to put pressure on "climate fiction" as it relates to work in postcolonial environmental humanities, then, I hoped to put pressure on existing definitions of terms like *migrant* and *refugee* in the context of climate change.

International human rights law does not currently recognize climate change as a legal justification for seeking asylum (Dawson 1),[4] even as the current so-called refugee crisis at the southern border of the United States is motivated not only by political instability and gang violence in the Northern Triangle of Central America, comprising Honduras, El Salvador, and Guatemala, but also by climate change. How might accounting for climate change within legal and activist representations of the "refugee" change our understanding of who migrants are and why they move? And where might "migration literature" (Ahmed; Burge) intersect with the category of climate fiction?

One important point of confluence is that both migration literature and climate fiction force questions of temporality. It is certainly possible to consider a collection of texts about migration across a wide span of time, but people move for specific reasons and within specific contexts. Similarly, although one can draw together a range of texts under the rubric of environmental fiction, climate fiction is about particular large-scale environmental changes caused and experienced by human and nonhuman agents. Climate fiction has too often located these shifts in the future by means of narratives about apocalyptic survivor stories in the Global North. But what happens when the frame shifts from the so-called receiver nations of global migration to center the already existing climate displacements in the Global South and for Indigenous communities around the world? As Ben put it, these texts offer conceptual frameworks for thinking about patterns of extractive violence and displacement spurred by colonialism and accelerated by climate change. For example, Dimaline's *The Marrow Thieves* adopts the near-future setting of much cli-fi but grounds that future in the centuries-long displacement of First Nations communities from their ancestral lands around Toronto into the cities and residential schools. Northward migration in the novel is a response not only to environmental destruction but also to genocidal white supremacy.

That question of temporality leads to crucial questions about differential impact. If it is overwhelmingly Indigenous and Global South communities that are already experiencing the most devastating effects of climate change, the most innovative interdisciplinary research points toward exponentially escalating harm within the next fifty years. Abrahm Lustgarten's excellent article "The Great Climate Migration," part of a collaboration between ProPublica and *The New York Times Magazine*, offers an overview of recent findings by research teams involved in large-scale modeling projects combining data on political stability, crop yields, changing temperatures, wages, and other factors. Researchers found that, even with the numerous variations produced by their ten billion data points,

> one trend is clear: Around the world, as people run short of food and abandon farms, they gravitate toward cities, which quickly grow overcrowded. It's in these cities, where waves of new people stretch infrastructure, resources and services to their limits, that . . . the most severe strains on society will unfold. Food has to be imported—stretching

reliance on already-struggling farms and increasing its cost. People will congregate in slums, with little water or electricity, where they are more vulnerable to flooding or other disasters. The slums fuel extremism and chaos.

In other words, they found a narrative. It is a narrative about migration, describing first internal migration (an earlier version of the modeling project estimated 143 million internal migrants in South Asia, Latin America, and sub-Saharan Africa alone) and then, as resources run out and political instability intensifies, international migration on a massive scale.

As scholars working at the nexus of postcolonial studies, migration, and social justice, we hope that teaching about climate fiction will entail a capacious and interdisciplinary grappling with this unfolding narrative. In what ways does contemporary climate fiction already engage with large-scale migration as a consequence of climate change, and in what ways do representations of environmental refugeeism in climate fiction confront or avoid the legacy of white supremacist colonialism that has already fueled displacement and flight? Climate fictions from the Global South and Indigenous communities provide important insights about the imbrications of racial, decolonial, and environmental justice; they should be included in syllabi not as simple additions but as part of a process of restructuring climate teaching to center those concerns. When taught in a way that engages their historical and geographic contexts, such texts provide new conceptual frameworks for the field as a whole, influencing the kinds of questions we ask, the communities we study, and the futures we imagine.

We do not attempt to offer definitive answers here but rather to encourage each other to do the difficult teaching work of holding on to more than one thing at once. As Ben found in their class, this can be a pedagogical challenge. How do we center climate change and migration without losing a sense of terminological precision and focus? Emily has similarly found teaching in this intersectional mode both vulnerable and deeply fulfilling: as a teacher, one knows less about more things but can model asking questions and pursuing research that connects struggles that have too often been presented as discrete from one another. Historicizing and connecting struggles for racial justice and climate change enables students to perceive and tell new stories about what has happened, what might happen, and what we can seek to actualize instead. As communities of readers, we see new stories about how we have already been connected and how we have already been implicated in the dispossession and suffering of others. Climate fiction's temporality points us forward; however, to avoid replicating the Euro-American centrism and racism that has produced climate injustice, just futures must emerge from a deeply grounded and intersectional sense of location and historicity.

Notes

1. In using the terms *Global South* and *Global North*, we do not intend to imply that these regions are homogeneous or that their structural differences are a result of geographic determinism. Instead, we recognize geopolitical patterns of uneven power distribution, which, among their many effects, lead to authors from colonized nations being taught less often in the American classroom than their more privileged counterparts.

2. Matthew Schneider-Mayerson argues that most influential American climate fictions portray "climatic destabilization primarily as a problem for white, wealthy, educated Americans and secondarily [gesture] toward its consequences for . . . the monolithic and flattened 'we' of *homo sapiens*" (945). Both modes ignore climate justice, from which perspective different humans have differentiated responsibilities and vulnerabilities. If American cli-fi lacks this framing, there is all the more reason to recenter Global South fictions.

3. Jennifer Wenzel offers the method-focused rubric of "reading for the planet"; Elizabeth DeLoughrey highlights allegory; and Upamanyu Pablo Mukherjee suggests "uneven form," for example.

4. Simon Behrman and Avidan Kent's collection traces the history and debates in international refugee law that have produced this current situation.

Works Cited

Ahmed, Dohra, editor. *The Penguin Book of Migration Literature: Departures, Arrivals, Generations, Returns.* Penguin Classics, 2020.

Behrman, Simon, and Avidan Kent, editors. *Climate Refugees: Beyond the Legal Impasse?* Routledge, 2018.

Bloom, Dan. *The Cli-Fi Report Global.* 2023, www.cli-fi.net/.

Burge, Amy. *What Can Literature Tell Us about Migration?* IRiS Working Paper Series 37, U of Birmingham, Institute for Research into Superdiversity, 2020.

Craps, Stef, and Rick Crownshaw. Introduction. *The Rising Tide of Climate Change Fiction*, special issue of *Studies in the Novel*, edited by Craps and Crownshaw, vol. 50, no. 1, 2018, pp. 1–8, https://doi.org/10.1353/sdn.2018.0000.

Dawson, Ashley. "The Right to Stay and the Right to Move." *Journal of Postcolonial Writing*, vol. 57, no. 1, 2021, pp. 1–3.

DeLoughrey, Elizabeth. *Allegories of the Anthropocene.* Duke UP, 2019.

Dimaline, Cherie. *The Marrow Thieves.* Dancing Cat Books, 2017.

Evans, Rebecca. "Fantastic Futures? Cli-Fi, Climate Justice, and Queer Futurity." *Resilience: A Journal of the Environmental Humanities*, vol. 4, nos. 2–3, 2017, pp. 94–110.

Ghosh, Amitav. *The Great Derangement: Climate Change and the Unthinkable.* U of Chicago P, 2016.

———. *The Hungry Tide.* First Mariner Books, 2006.

Jetñil-Kijiner, Kathy. *Iep Jāltok: Poems from a Marshallese Daughter.* U of Arizona P, 2017.

Johns-Putra, Adeline. *Climate Change and the Contemporary Novel.* Cambridge UP, 2019.

LeMenager, Stephanie. "Climate Change and the Struggle for Genre." *Anthropocene Reading: Literary History in Geologic Times,* edited by Tobias Menely and Jesse Oak Taylor, Penn State UP, 2017, pp. 220–38.

Lustgarten, Abrahm. "The Great Climate Migration." Photographs by Meridith Kohut, *The New York Times Magazine,* 23 July 2020, nytimes.com/interactive/2020/07/23/magazine/climate-migration.html.

Marshall, Kate. "What Are the Novels of the Anthropocene? American Fiction in Geological Time." *American Literary History,* vol. 27, no. 3, pp. 523–38.

Mukherjee, Upamanyu Pablo. *Postcolonial Environments: Nature, Culture and the Contemporary Indian Novel in English.* Palgrave Macmillan, 2010.

Satgar, Vishwas, and Jane Cherry. "Climate and Food Inequality: The South African Food Sovereignty Campaign Response." *Globalizations,* vol. 17, no. 2, 2020, pp. 317–37, https://doi.org/10.1080/14747731.2019.1652467.

Schneider-Mayerson, Matthew. "Whose Odds? The Absence of Climate Justice in American Climate Fiction Novels." *ISLE: Interdisciplinary Studies in Literature and Environment,* vol. 26, no. 4, 2019, pp. 944–67, https://doi-org.udel.idm.oclc.org/10.1093/isle/isz081.

Trexler, Adam. *Anthropocene Fictions: The Novel in a Time of Climate Change.* U of Virginia P, 2015.

Wenzel, Jennifer. *The Disposition of Nature: Environmental Crisis and World Literature.* Fordham UP, 2020.

Part V

Assignments

Orchid Tierney

Tuning In to Climate Change: Podcasts in the Classroom

Climate Emergencies is an English literature course open to first- and second-year students at Kenyon College, a small liberal arts college in rural Ohio. In brief, the course explores anglophone literature from 1945 to the present through the lens of climate change and public scholarship. Students learn how writers have theorized and represented environmental emergency with inventive strategies that foreground racial, economic, and environmental justice. Throughout the semester we attend to a variety of questions, with the following taking special precedence: What is climate justice? What does it mean to witness—and resist—extinction? And, lastly, what is hope in our prolonged emergency?

The course stages a variety of Indigenous, anti-racist, and postcolonial analyses on sea-level rise, global warming, and climate disruption with respect to poetry, fiction, and nonfiction from Pacific Island nations, including the Marshall Islands, Guam, and Aotearoa New Zealand, although I also include select readings from the United States, Australia, and India. Overall, these readings foster students' critical sensitivity toward the intersections of race, empire, and climate justice as well as the uneven distribution of environmental risks in local and global contexts. Typical texts include Keri Hulme's short story "Floating Words," David Eggleton's *Time of the Icebergs*, John Kinsella's *Firebreaks*, Kathy Jetñil-Kijiner's *Iep Jāltok: Poems from a Marshallese Daughter*, Craig Santos Perez's "Praise Poem for Oceania," and Amitav Ghosh's *The Great Derangement: Climate Change and the Unthinkable* in addition to examples of public writing such as James Bradley's "Ocean and an Instant" and Naomi Klein's "Climate Rage." These assigned texts introduce the class to contemporary social and political infrastructures and systems that reproduce racial, gender, and economic disparities while demonstrating in some manner a socially inflected response to the place-specific challenges of climate change.

While the Pacific Ocean may feel remote to students in the rural Midwest, the issues of climate change are as resonant in Aotearoa New Zealand as they are in Ohio. With this in mind, the learning goals of this course are fairly straightforward. First, the course pushes students to rethink notions of agency, resilience, and global citizenship in the context of place specificities while recognizing the cultural discourses on climate change that inform their localized positionalities and group identifications. This task is no easy feat, but encouraging students to think globally by engaging locally relates to the second learning goal of the course: students must collaborate in pairs to produce a ten-minute research-based podcast on a subject of their choice that engages the themes of the syllabus and includes interviews with members of the local community. Broadcast media is relentless with examples of dire climate catastrophes and misinformation, which leave students in this class reflecting on their shared sense of grief over a world they are inheriting and their desire to discuss climate change with others who may be skeptical. They feel an urgent need to respond concretely to climate change, and the podcast is a small exercise in enacting their agency and in communicating the language of climate justice to a community or group outside their own. It is also a means for students to support community outreach by interviewing a member of the college, town, or county and by testing alternative modes of information distribution on an environmental topic. Podcasts won't prevent climate change, but they do enable students to participate in public conversations about it. In other words, podcasts are sonic forms of civic commitment.

Why Podcasts?

Classified by the college as an entry-level community-engaged learning course, Climate Emergencies is structured around an interview-based podcast.[1] My rationale for this assignment is based on the idea that podcasts are knowledge work and that undergraduate students can imagine themselves as public scholars. Further, Leigh A. Jones proposes that podcasts underscore the "potential for exploring the aural mode of communication in service of the written" (77), and to this end, the course combines podcasting with written forms of nonacademic scholarship—namely, the personal essay. Learning to communicate difficult ideas in a nonacademic context and across group identifications involves unpacking the podcast genre. The process of collaborating on, researching, and executing a podcast requires group compromises, being familiar with one's audience, and attending to the rhetoric of climate communication. Consequently, we spend a substantial portion of class time early in the semester discussing precision in oral communication, accessible transcriptions, audience, and audio rhetoric. For example, we unpack the nuances of climate terminology (for example, *climate change, global warming, the weather, environmental justice, resilience,* and *sustainability*), public confusion about climate modeling, ideological differences, and the challenges of communicating science and abstract ideas through storytelling.[2] Whether podcasts are truly effective tools to combat negative climate beliefs is moot, but as Mark J. W. Lee, Catherine McLoughlin, and Anthon Chan have argued, podcasts do have "knowledge-creation value" and can be used as "a vehicle for disseminating learner-generated content" (504). In this way, the podcast assignment supports the course rationale of encouraging students to think of themselves as public scholars who can communicate climate discourse to a nonacademic audience.

Assignments

Generally I assume that few—if any—students have any prior exposure to podcasting or audio editing, and I distribute audio skills evenly across the various pairings (which are a combination of student choice and my arrangement). For this reason, lesson plans are designed to introduce the class to the fundamentals of storytelling, audio capture, interviewing, and editing. I also touch on the basics of e-mail etiquette (how to request an interview, for example), sound and music copyright, and media distribution and interview consent forms. Students must submit these forms with their final podcasts. I don't discuss these lessons in detail here, but I cover the most significant assignment steps below.

It goes without saying that podcasts require substantial preparation. Indeed, this course only has one real assignment—the podcast—but the steps involved, including research, scripting, and editing, require ladder-like tasks that build up learner confidence and media literacy to a point where pairs can interview a member of the public and edit the content into a reasonably polished podcast. The smaller assignments, which are all graded, also serve to lower the stakes of the podcast (which is worth twenty percent of the final grade). Recognizing, too, that sometimes collaborations go awry, interviews fall flat, audio technology fails, or a student experiences a significant life event, keeping the podcast low-stakes allows students to exit the course having learned various podcasting techniques while giving themselves the freedom to make mistakes or drop an exercise without worrying about their final grades.

The assignments are broken down into the following tasks: a listening exercise, a personal essay, a proposal, a podcast script, a podcast, and a reflection paper. The proposal, script, and podcast are the only pair-produced assignments. As a way to introduce the class to audio techniques, we start with a listening exercise. Students select from a preestablished list of climate-oriented podcasts before responding to the following questions:

> How does the podcast begin? Are any advertisements included? Who sponsors the podcast?
>
> How does the narrator introduce their topic?
>
> Does the podcast deliver an argument, a story, or an interview? How does the narrator transition between the various elements? How is the podcast structured?
>
> How do music and sound effects shape and support the podcast's topic? Do you notice any pauses? If so, how are they used?
>
> Describe the narration or the speakers' delivery. Is it scripted or casual?
>
> Who is the intended audience? How do you know?[3]

The class presents their findings orally, and together we crowdsource similarities and differences between the podcasts. The intention behind this exercise is to increase students' awareness of sonic rhetoric and their understanding of orality and sound effects as strategies for engaging an audience through storytelling or a persuasive argument. In their end-of-semester reflection paper, one student, Shea, mentioned having found this exercise particularly helpful and having "used a fair amount of what I learned from it for our own podcast." Much of this exercise is geared toward curating a collective group awareness of the relationships between sound and narrative arcs, transitions, and climate discourse that can effectively engage a general audience.

As I mentioned earlier, podcasts can be combined with other forms of scholarship and writing. In this class, students write a personal essay to help them practice storytelling and translating concepts gleaned from the assigned reading to a general readership. They begin by reading James Bradley's "Ocean and an Instant," an essay on familial memory and catastrophe, which serves as a reference for a guided discussion on how to weave structure, language, and research together into a compelling arc. Students then produce their own essays of four to five pages in which they attempt to concretize the impacts of climate change through their own personal experiences.[4] Sam, for example, was interested in illustrating his shock at having been caught in an unexpected hailstorm while driving with his friend. His condensed descriptions wonderfully underscore the suddenness of nature inflicting itself on human vulnerability: "A large ball of rage planted itself into the windshield, completely shattering it," he writes. "I immediately snapped back to reality, glass spread across my lap." The challenge here is to force students to narrow the scope of their topic. They don't need to convey everything about their experiences of climate change. Rather, their mission is to crystallize their diverse ideas through a single encounter, event, or moment as expertly as Sam did.

The personal essay is an early exercise that invites students to think about how they might translate abstract ideas in an audio and vernacular context. The next steps encourage them to collate their listening and creative skills into their podcasts. First, each pair submits a proposal (also called the "pitch") on a topic they want to pursue. The pitch is roughly two to three pages and helps the teams develop their initial ideas and explain why their topic is compelling to a listener. This working document includes resources I provide, such as rudimentary research on their topic, a list of possible interviewees, and suggestions for music and sound effects from public domain sources.[5] Most pairs revise their proposal throughout the semester as they develop different research strands.

Similarly, the next exercise, the script, is not a fixed document; it does, however, need to include a breakdown of interview questions, background information on the interviewee, a possible narrative arc for the podcast, and any key points students want to cover. (The teams can script their dialogue too, although their script will depend on the kind of audience they are envisioning). The script gives the class an opportunity to practice, as the cliché goes, "writing for the ear." As the class is preparing for their community interaction at this stage, we hold mock interviews so the pairs can practice asking questions that will elicit responses, test technologies for capturing audio, and develop their familiarity with our editing software of choice, *Audacity*. (I permit students to use any technology or platform for recording or editing with which they are already familiar. These programs run the gamut from *Zencastr* to *Zoom*, *Google Meet*, *Descript*, *Adobe Audition*, and *Garage Band*.[6]) Yet the most important lesson I stress at this time is the "five *P*'s": Perfect Planning Prevents Poor Performance. (Along these lines, one student, Lynn, cautioned future students in her final reflection paper not to "underestimate the amount of time it takes to edit audio.")

Students have submitted a strong range of podcasts discussing the effects of climate change on the café and coffee production industries, programs that preserve the Ohioan landscape, the presence of climate politics on campus, and Chang-rae Lee's *On Such a Full Sea*. Interviewees have included farmers, baristas, a director of a local preserve, experts on solar panels, engineers developing community climate resilience endeavors, and students from diverse backgrounds reflecting on the relationships between art and climate change. Once they have submitted their consent forms and audio, students submit individually written reflection papers in which they describe their experiences of making and editing the podcast. Here they consider what they've learned, where they might improve, how they understand their interactions with the community, what challenged them, and what advice they would give to future podcasters. Overall, students have commented on how invaluable they find the collaboration process. Salome, for example, notes, "[W]hen I work in group projects, I really like to control the situation and know exactly what the project is going to manifest itself into. . . . I learned to trust my partner and know that whatever he comes up with will meet my expectations." Alexia, who coproduced a podcast on solar panels in Knox County, agrees: "I learned from our collaboration how to create something with someone else but also make it my own." Not all collaborations will be productive and smooth, but embracing collaboration as a mode of learning through generosity and disagreement is an essential ingredient for a successful community partnership.

Podcasts as Student-Driven Scholarship

I have defined podcasts as knowledge work, and many students discover that this assignment gives them permission to expand on their classroom discussions and their readings of the assigned texts on climate change. Tatumn, who expressed some initial hesitation over the podcast assignment, came to enjoy her collaboration as a means of furthering her understanding of climate resilience within the community. "As a class, we had talked about what resilience meant and what things we have all done to be resilient," she writes. "Though [an] action may seem small, it can make a large impact, especially if many people do it." Tatumn and her partner subsequently structured their narration for a teen audience, using the podcast to reflect both on their personal experiences and on state climate policies. As Tatumn suggests, "[H]aving a podcast where a college student can explain these things in a way that makes sense is much more appealing to teens wanting to learn what they can do." Tatumn and her classmates are motivated to reach groups on the periphery of the climate change discourse. Similarly, Alexia notes, "[T]he biggest challenge we as environmentalists face is getting people to care." Yet during the course of the podcast assignment, she was delighted to learn that

> people are starting to understand the importance of environmentalism. Even though Knox county is a relatively conservative area, they are seeing how their environment has changed and want to help. The creation of the Climate Change Coalition of Knox County gives me hope that attitudes are changing. Not only are people, specifically in this conservative rural place, confronting that there is a problem, but they are acting upon it.

Alexia's comment underscores how community engagement can foster understanding of a community that is mobilizing its resources to respond to climate change. The interview assignment offers a civic interaction with local participants, who can position themselves as experts, co-collaborators, and co-learners in dialogue with students rather than as resources to be tapped into by educational programs. This kind of community engagement enacts what David D. Cooper calls a "learning in the plural"—that is, "learning in community and community in learning" (xx)—which supports students in reconsidering the nature of authority and knowledge production in climate discourse. And this learning outcome is not lost on students. As a result of coproducing a podcast, Tatumn is positive about the future. "My outlook on many things [has] changed," she writes. "I am excited to take this new knowledge with me."

Notes

1. Entry-level, in this case, means the community engagement in the class is limited to a single interview interaction.

2. For these discussions, I draw on Shome and Marx.

3. The list of podcasts includes Kuntz; Husain; Gill et al.; Gill and Laungani; and Caners and Andrews, among others.

4. Another example I have used for this exercise is the Climate Stories Project (climatestoriesproject.org/).

5. I share a list of interviewees based on the podcast ideas the student teams have proposed at an earlier conference. Knox County is a small community, and I assign interviewees to teams to ensure even distribution. The college's Office for Community Partnerships provides the pool of interviewees, who have already participated in a community-engaged course at the college. Teams select three possible interviewees from this pool. I may need to assign a popular interviewee to a team, but I allow students to interview one of their three choices. Students may also identify new community partners to interview, provided they seek permission from the instructor before making contact. (As an alternative, students can propose to interview a subject from their hometown.) Additionally, I dedicate at least one class session to copyright and public domain music and go through the basics of working with consent forms.

6. I am grateful for the support of my colleagues Joe Murphy, Alex Alderman, and Ashley Butler, who visit my class from the Center for Innovative Pedagogy to instruct students on audio equipment and editing.

Works Cited

Bradley, James. "An Ocean and an Instant." *Sydney Review of Books*, 21 Aug. 2018, sydneyreviewofbooks.com/essay/an-ocean-and-an-instant/.

Caners, Kevin, and Tony Andrews. "Iceland and the Surprising Science of Turning CO_2 into Rock." *The Elephant*, 20 July 2019, www.elephantpodcast.org/episodes/iceland-and-the-surprising-science-of-turning-co2-into-rock.

Cooper, David D. *Learning in the Plural: Essays on the Humanities and Public Life.* Michigan State UP, 2014.

Eggleton, David. *Time of the Icebergs.* Otago UP, 2010.

Ghosh, Amitav. *The Great Derangement: Climate Change and the Unthinkable.* U of Chicago P, 2016.

Gill, Jacquelyn, and Ramesh Laungani. "Telling Human Stories." *Warm Regards*, 27 July 2020, warmregardspodcast.com/episodes/telling-human-stories-s1!8b256.

Gill, Jacquelyn, et al. "The Dangers of Doing Science in the Field." *Warm Regards*, 8 May 2019, warmregardspodcast.com/episodes/the-dangers-of-doing-science-in-the-field-s1!5ac29.

Hulme, Keri. "Floating Words." *Stonefish*, by Hulme, Huia, 2004, pp. 5–19.

Husain, Ellen. "Defenders of the Reef." *Costing the Earth*, BBC Radio Four, 27 Feb. 2018, www.bbc.co.uk/sounds/play/b09snj90.

Jetñil-Kijiner, Kathy. *Iep Jāltok: Poems from a Marshallese Daughter.* U of Arizona P, 2017.

Jones, Leigh A. "Podcasting and Performativity: Multimodal Invention in an Advanced Writing Class." *Composition Studies*, vol. 38, no. 2, 2010, pp. 75–91.

Kinsella, John. *Firebreaks.* W. W. Norton, 2016.

Klein, Naomi. "Climate Rage." *Rolling Stone,* 12 Nov. 2009, www.rollingstone
.com/politics/politics-news/climate-rage-193377/.

Kuntz, Lauren. "The Making of a Climate Scientist with MIT Terrascope
Alumna Lauren Kuntz." *Climate Conversations: A Climate Change Podcast,*
7 Nov. 2018, climate.mit.edu/podcasts/climate-conversations-s3e6-making
-climate-scientist-mit-terrascope-alumna-lauren-kuntz.

Lee, Mark J. W., et al. "Talk the Talk: Learner-Generated Podcasts as Catalysts
for Knowledge Creation." *British Journal of Educational Technology,* vol. 39,
no. 3, 2008, pp. 501–21.

Perez, Craig Santos. "Praise Poem for Oceania." *Habitat Threshold,* by Perez,
Omnidawn Publishing, 2020, pp. 66–72.

Shome, Debika, and Sabine M. Marx. *The Psychology of Climate Change
Communication: A Guide for Scientists, Journalists, Educators, Political
Aides, and the Interested Public.* Center for Research on Environmental
Decisions, Columbia U, 2009, https://doi.org/10.7916/d8-byzb-0s23.

Melissa Anderson

The Literature of Climate Change
and Information Literacy Instruction

Traditionally, information literacy instruction in literature courses has been limited to bibliographic instruction—searching databases, using the library catalog, and creating accurate citations—whereas many of the higher-level critical thinking skills associated with information literacy have been left to other courses. However, since the adoption of the *Framework for Information Literacy for Higher Education* by the Association of College and Research Libraries in 2016, the focus in information literacy instruction has been shifting. Instead of concentrating on purely bibliographic instruction, instructors and academic librarians have been teaching more sophisticated concepts, such as the complicated construction of authority in writing, the significance of format and genre in our perception of a text, and the role of creativity in the production of knowledge—all ideas that align exceptionally well with the teaching of literature. Climate change as a topic lends itself naturally to information literacy instruction, since information on the subject has been debated, analyzed, and disseminated across multiple platforms and in multiple genres. The study of the literature of climate change expands the existing relationship between climate change and information literacy into an opportunity to teach a variety of transferable

246

critical thinking skills integral both to literary study and to college research and writing. Using current research in information literacy instruction and experience teaching writing and literature, this essay explores how the information literacy concepts of authority and format can be integrated in the study of climate change literature before presenting a sample assignment created to teach these concepts in conjunction with analysis of Barbara Kingsolver's *Flight Behavior.*

Threshold Concepts and Information Literacy Instruction

The *Framework for Information Literacy for Higher Education,* since 2016 the guiding document for teaching information literacy at the college level, is structured around the idea of threshold concepts, first introduced by Jan Meyer and Ray Land in 2003. Threshold concepts are transformative ideas that allow students to make new connections and begin to think like disciplinary experts; linked to David Perkins's notion of "troublesome knowledge," they are by nature integral to the building of new knowledge but also difficult to grasp (Meyer and Land 1). The *Framework* posits a number of threshold concepts as the basis for developing information literacy; these concepts not only allow students to improve their skills in searching for and evaluating information but also encourage students to think about their own role in the production and consumption of information, to contextualize notions of authority, and to grapple with epistemology. Building on the work that produced the *Framework,* Amy Hofer, Sylvia Lin Hanick, and Lori Townsend have further developed threshold concepts that present opportunities for interdisciplinary explorations of information literacy, and two of these—format and authority—are particularly relevant to the study of climate change literature.

Format

The threshold concept of format appears deceptively simple: information is created by a process, and the products of that process are accessed in specific formats, such as speeches, articles, books, tweets, videos, and so on.[1] However, embedded within this concept are the ideas that formats have recognizable characteristics and communicative purposes and that those reading, hearing, seeing, or otherwise engaging with a format have expectations regarding what that format will contain, how it will be presented, and the process used to create it. Recognizing these embedded ideas raises epistemological questions about specific formats (Hofer et al.). Instructors can use discussions of genre and genre theory, in and of themselves useful in the study of literature, to help students grasp the concept of format. Hofer, Lin Hanick, and Townsend explain that in discussions of format, a genre is "a text with a purpose" and that text, in whatever format, whether it be a scholarly journal article, a popular book, or a blog post, has rules and conventions that govern its structure and content (85). "These rules or conventions," they write, "help people quickly make sense of what they are reading, hearing, or seeing" (86). Information about climate change exists in myriad formats and genres that can be examined, compared, and evaluated both within a specific piece of climate change literature and also in the larger information landscape. What is more, while thinking about what the format of a climate text signifies, students will also begin to delve into other information literacy threshold concepts, such as authority, since the expected credibility of various genres is also at play in an examination of format. In the example assignment below, students examine a variety of information formats from their own Internet searches and also examine the different formats used for scientific information in Kingsolver's *Flight Behavior*. Kingsolver's novel is a particularly useful vehicle for information literacy instruction because the narrative involves characters struggling with information literacy concepts and because the novel itself is in dialogue with scientific information about climate change; students can therefore explore the practice of information literacy on two levels.

As has been noted, *Flight Behavior* evokes both "scientific and experiential epistemologies" (Houser 111), and characters engage with information sources as varied as websites, television news stories, research data, lectures, sermons, and more, making the text ripe for a discussion of format. Early in the novel, Dellarobia Turnbow uses *Wikipedia* to find information on monarch butterflies, which she then shares with Ovid Byron over dinner. For her initial purposes of gaining an overview of basic information, *Wikipedia* is a credible scientific information format. When Ovid begins talking about the butterflies, however, Dellarobia quickly realizes that there are other information formats that, by virtue of their process of creation and scientific authority, provide much more comprehensive information:

> "So, you do what, experiments, or observations? And write up what you find out?"
>
> He nodded. "A dissertation, articles, a couple of books. All on the monarch." (Kingsolver 121)

After this interchange, she thinks to herself, "So there were worse things than feeding meat loaf to a vegetarian. Like blabbing wiki-facts to the person who probably discovered them in the first place" (121). This passage lends itself beautifully to a discussion of how scientific information is created and disseminated in formats as various as dissertations, journal articles, and *Wikipedia* entries. It also can be used to talk about what our expectations are for each of those formats—what they look like, what types of information are included in each one, where to find them, when one would use them, and whether or not they are appropriate sources in different social and professional situations.

Authority

In academic courses, the concept of authority is often invoked when instructors tell students they must use peer-reviewed sources or library databases. The *Framework*'s assertion that "authority is constructed and contextual," however, asks students to wrestle with the idea that rather than being inherent in a particular information format, authority is "conferred by the information user upon the information source" (Hofer et al. 56). What is more, it is something that is "built up over time" and "is based in social relationships and communities of practice" (57, 58). Instructors have incorporated this idea into their own approach to research almost unconsciously; they know why the work of some experts is given more weight than the work of others on specific topics and how what counts as an expert opinion changes in different social and professional contexts. Students have a harder time articulating what counts as expertise in different situations and may rely on format as the sole determinant of authority.

It can also be difficult for students to understand that the value of any marker of authority—academic credentials, job title, publication history, and even format itself—can shift in different contexts. In the novel, Dellarobia witnesses this shift when she discusses climate change with Cub. She notes that, although Ovid, a scientist, and Bobby Ogle, a preacher, share a similar knowledgeable and forthright demeanor, the townspeople dismiss the scientist's views where they easily accept the views of the preacher. "He's not from here, that's the thing," Cub explains, demonstrating his understanding of the importance of a known and local identity in establishing authority in his community (Kingsolver 257). Dellarobia then tries to explicate the climate science behind the butterfly migration, but Cub refuses to accept it, instead parroting disparaging retorts he has heard from the local radio personality. "Why would we believe Johnny Midgeon about something scientific, but not the scientists?" Dellarobia asks Cub. "Johnny Midgeon gives the weather report," he responds (261). In the context of rural Tennessee, the scientific authority of Ovid and his team of researchers pales in comparison with that of the local authority on the weather. This is not to say that Cub and those sharing his views are "right" to believe the pronouncements of a shock jock over the explanation of a biologist, because the question is not actually about what is "right" or "true"; it is about what constitutes authority for a specific audience in a specific scenario. The idea that authority is separate from truth has been the subject of much debate about the *Framework*, demonstrating that authority is a troublesome concept indeed (see, for example, Rinne; Baer). Examining authority in *Flight Behavior*, however, allows students to see how authority is constructed in a very specific cultural context and can encourage them to examine the construction of authority in other contexts, both academic and personal.

The Assignment

Created for an interdisciplinary course on climate change in fact and fiction, the assignment (see appendix) was designed to help students practice evaluating different information formats and types of scientific authority while also practicing close reading and analysis of a literary text. The assignment is structured by Mary-Ann Winkelmes's principles of transparent assignment design, which have been shown to facilitate metacognition—itself an important part of information literacy—and support greater student persistence and retention, particularly in underserved student populations (32). As scaffolding for a later paper assignment on the representation of science and climate change in *Flight Behavior*, students are asked first to find and evaluate articles about climate change while thinking about both format and authority and then to look at passages in *Flight Behavior* that also explore different types of information formats and scientific authorities. After assignments are turned in, the articles and passages found by students form the basis of a rich class discussion about information formats, scientific authority, and the role of bias in our interpretation of scientific information. The end goal is that students practice recognizing and evaluating format and authority in a way that will help them become more adept at assessing them in all contexts—in literature courses, in professional contexts, and in their daily lives.

Note

1. Hofer, Lin Hanick, and Townsend use genre theory to explain the way format works as a threshold concept. Borrowing terms from the field of rhetoric, they explain that formats, in the specific way that they are understood by librarians, work like genres because they come with associations about structure, content, and social context.

Works Cited

Baer, Andrea. "It's All Relative? Post-Truth Rhetoric, Relativism, and Teaching on 'Authority as Constructed and Contextual.'" *College and Research Libraries News*, vol. 79, no. 2, 2018, pp. 72–97, https://doi.org/10.5860/crln.79.2.72.

Framework for Information Literacy for Higher Education. Association of College and Research Libraries, 2016, www.ala.org/acrl/standards/ilframework.

Hofer, Amy, et al. *Transforming Information Literacy Instruction: Threshold Concepts in Theory and Practice.* Libraries Unlimited, 2019.

Houser, Heather. "Knowledge Work and the Commons in Barbara Kingsolver's and Ann Pancake's Appalachia." *Modern Fiction Studies*, vol. 63, no. 1, 2018.

Kingsolver, Barbara. *Flight Behavior.* HarperCollins Publishers, 2012.

Meyer, Jan, and Ray Land. *Threshold Concepts and Troublesome Knowledge: Linkages to Ways of Thinking and Practising within the Disciplines.* ETL Project, Occasional Report 4, Economic and Social Research Council, 2003, www.etl.tla.ed.ac.uk/docs/ETLreport4.pdf.

Perkins, David. "Constructivism and Troublesome Knowledge." *Overcoming Barriers to Student Understanding: Threshold Concepts and Troublesome Knowledge*, edited by Jan H. F. Meyer and Ray Land, Routledge, 2006, pp. 33–47.

Rinne, Nathan Aaron. "The New Framework: A Truth-Less Construction Just Waiting to Be Scrapped?" *Reference Services Review*, vol. 45, no. 1, Jan. 2017, pp. 54–66, https://doi.org/10.1108/RSR-06-2016-0039.

Winkelmes, Mary-Ann, et al. "A Teaching Intervention That Increases Underserved College Students' Success." *Peer Review*, vol. 18, nos. 1–2, 2016, pp. 31–36, www.aacu.org/peerreview/2016/winter-spring/Winkelmes.

Appendix: Assignment on Establishing Scientific Authority in Information Sources

Purpose

Alongside our discussions about what gives science "authority" in contemporary society, we have been discussing the different kinds of "science" in *Flight Behavior* and Dellarobia's complex relationship to science and scientists. One of our course goals is to learn to recognize the role scientific discourse plays in culture more generally and in literature in particular. Therefore, in this assignment you will analyze contemporary information sources on a scientific topic and evaluate their "authority" for particular audiences and contexts. This exercise will contribute to your overall ability to read and interpret scientific information sources in this class and beyond. In addition, you will use your close reading skills to analyze passages in *Flight Behavior* to see how Kingsolver establishes or problematizes the authority of the science in the novel. This assignment will prepare you for the next assignment, which is a paper on science and climate change in the novel.

Skills

The purpose of this assignment is to help you practice the following skills that you will use in other courses and in your professional and personal life beyond school:

1. Recognizing how authors establish authority for scientific information in various types of information sources
2. Evaluating the authority of scientific information in different information formats
3. Analyzing how scientific terms, ideas, and themes work in a literary text

Knowledge

This assignment will also help you become familiar with content knowledge in the following areas:

1. How and where people encounter scientific information
2. The expectations people have of different information formats
3. How scientific knowledge is created and disseminated

Task

1. Find three articles online that were published in the past two years on a topic related to climate change. They should come from three different sources, but all should be written for a general audience (not as peer-reviewed articles for scientific journals). These articles may be in a variety of formats, such as newspapers, blogs, and government websites. Provide a complete citation using MLA style for each article you find.

2. State the format (blog post, newspaper article, etc.) of each information source and answer the questions below. For each article, describe where the scientific authority comes from and explain why those features are considered authoritative. Determine the bias, if any, for each article and information source by analyzing the purpose of the article and the publication or website, the context for the article, and any relevant information you have about the author. . Explain why you think bias exists or doesn't exist for each article.

3. Choose two passages from *Flight Behavior* and describe what kind of scientific authority is established in each passage. There are suggestions

for passages on the assignment template, but you can choose your own passages if you like. Remember that *scientific* can be a subjective term, so you will need to describe (a) what kind of science (biology, entomology, animal husbandry, climatology, etc.) is at play in the passage, (b) what constitutes authority for that type of science, and (c) who recognizes that type of authority as authoritative. Explain whether or not all the characters recognize or validate that authority and what might call it into question for Dellarobia, for other characters, or for the reader.

4. Explain how each passage fits into the overall narrative of climate change in the novel—does it support it? complicate it? refute it? Do you think the reader experiences the science in the passage the same way that Dellarobia does, or is the reader's perception of the larger scientific narrative different from hers at this point in the text?

Criteria for Success

1. A successful assignment will be turned in on time.

2. Each outside article will meet the criteria and will have a full citation in MLA style with a link to where the article was found online (you do not need to attach the articles themselves).

3. All questions will be answered completely and will include an explanation of why you answered as you did.

4. Two different passages from *Flight Behavior* will be chosen, all questions will be answered, and the analysis of scientific authority will include quotations, terms, or themes from the passage.

5. The last question, which asks you to make a claim about how the passage relates to the rest of the novel, will include evidence from the text that supports the claim made.

After the assignment is graded, we will go over some of the online articles found by students in class and also discuss passages used by students for the assignment, which will prepare you for the paper assignment.

Assignment Template/Worksheet

Part 1: Science Articles

For each of the three articles you find, provide a full MLA citation with a link and answer the following questions:

1. What type of information format is this article (newspaper article, blog post, government website, etc.)?

2. What type of authority for scientific information is provided in this article (credentials, studies, etc.)?

3. Do you think this article exhibits bias? Why or why not?

Part 2: *Flight Behavior* **Textual Analysis**

For each of two passages presenting some type of science in *Flight Behavior,* give chapter and page numbers for the passage and answer the questions below. You may choose your own passages or use passages from the following list:

Chapter 5, pages 116–22
Chapter 6, pages 140–45
Chapter 9, pages 216–21
Chapter 10, pages 256–61
Chapter 12, pages 330–36
Chapter 12, pages 353–59
Chapter 13, pages 394–98
Chapter 14 pages 414–19

1. What type of science is seen in this passage? Is the authority granted to it based on theory, knowledge, experimentation, or something else? Who is practicing or presenting it?
2. Who in the novel, and in the culture at large, recognizes this type of authority for science?
3. Are there individuals or groups who do not recognize this type of authority as authoritative?
4. Does the reader experience the science in this passage differently from the characters in the novel, or in the same way? Why?
5. How does this passage fit into the overall narrative about climate change in the novel? Does it support it or question it? How?

Barbara Leckie

Noticing, Time, and Angling:
A Climate Change Syllabus

I begin my first-year seminar on climate change and the humanities by showing Hannah Gadsby's stand-up comedy routine *Nanette*. The students are perplexed. Why are we watching comedy, and what does this show about gender identity have to do with climate change? I explain that Gadsby zeroes in beautifully on several points that are important to our topic: that form is part of the story; that it is tough to find the words, concepts, and tone to convey difficult material; and that we can laugh. *Nanette* lets us discuss how we might represent climate change from a range of perspectives, and it helps students see that climate change is not only about science and information but also about emotion, affect, stories, and how those stories are told. As Gadsby points out, "[W]e learn from the part of the story we focus on" (00:41:08).[1]

This essay is composed of two parts. In the first half, I offer an overview of my course in relation to its focus on the arts of noticing, time, and angling as well as the three principles that underpin my approach: interdisciplinarity, collaboration, and conversation. In the second half, I elaborate on three teaching aids that have been successful in realizing these course principles: distribution of expertise, keyword work groups, and a one-day student-curated syllabus. I also demonstrate the many ways, both obvious and unexpected, that these activities and assignments foster interdisciplinarity, collaboration, and conversation. My goal is not to teach climate change per se but to teach the forms that climate change takes in a range of cultural works and to encourage modes of response adequate to the complexity and urgency of the climate crisis.[2]

Course Overview

I organize my course around what I call the arts of noticing, time, and angling. They are arts because they are practiced activities that are experienced differently by each student. At the outset, I use the course syllabus to illustrate these points. The syllabus introduces climate change through the tripart lens of noticing, time, and angling, and it accordingly shapes how students notice the material outlined. While I try to offer a range of perspectives and to illustrate the ways in which climate justice is inseparable from racial justice and questions of equity, I also note that the syllabus is necessarily limited and encourage students to discuss its gaps and occlusions. As it moves through the semester week by week, the syllabus also reinforces the time frame of the semester and the time slot of the class as a component of one's day. I note that although these approaches seem obvious and inevitable, they are not. During the pandemic, for example, my class had one synchronous *Zoom* class and one asynchronous class; the latter offered a different way of organizing the time of the class or, in the terms I have introduced, a different art of time. As this volume attests, not all climate change classes are alike, and all instructors shape how students will notice the material through, among other things, their course design.

I present climate change as a crisis of noticing and a crisis of time and suggest that literature has a unique capacity to address both. We often don't know how to notice climate change, for example, in ways that generate effective action. We also often don't know how to comprehend the demands that climate change places on conventional timescales. It is difficult to discuss the different temporalities climate change straddles, from the idea of deep time to the delayed release of carbon emissions, the urgency of the present moment, questions of futurity, and overlapping temporalities in general.[3] These challenges invite students to think about time in fresh ways beyond the linear time frames of past, present, and future to which they're accustomed.

In the context of both noticing and time, I seek to expand the range of things to which students pay attention and the modes through which they do so. We pay attention not only to what we notice but to how we notice; that is, we notice noticing. Similarly, we often experience time without thinking about how it shapes our experience. In one exercise, I ask students to select a space measuring three feet square and return to it each week to record what they notice in different media of their choice. Do they look up, look down? Do they imagine what the ant sees, what the bird sees? What do they smell, touch, hear? In a second weekly exercise, I ask students to explore time from as many angles as they can imagine. What does it feel like to spend an hour or an afternoon or a day without consulting any time device? Conversely, what happens if they record every time they look at a time device? What does it feel like to wake up early? to stay up late? to spend an allotted time period on social media and then to spend the same amount of time in a dark, quiet space?

The different perspectives that emerge in these exercises—what gets noticed and what does not, and how time assumptions inform what and how we comprehend—are part of what I call the art of angles and angling. I seek to demonstrate, through the students' experience and the material we address, that there are many different angles on climate change and that these angles are part of the story, just as comedy is part of the story that Gadsby tells. Different angles tell different stories. My goal, then, is twofold: to generate as many angles as we can on climate change and to make the angles themselves visible. This goal also embraces, albeit indirectly, the politics of climate change by drawing students' attention to how, with shifting angles, shifting approaches emerge. From a capitalist angle, for example, the response to climate change will often advocate for more responsible development as outlined in the United Nations' seventeen goals for sustainable development. From a socialist angle, the response might instead involve a rethinking and revision of social structures in general. From a Green New Deal angle, the response might be an attempt to blend responsible economic development with some social restructuring but not to fundamentally alter existing norms. In each case, the words, images, rhetoric, and stories mobilized to convey and promote one's position will be intimately bound up with how that position is understood and acted on.

In highlighting the arts of noticing, time, and angling, I have found several books especially helpful. Anna Lowenhaupt Tsing's *The Mushroom at the End of the World* at once discusses the matsutake mushroom and its flourishing in ruined environments, the temporalities of capitalism, and the many different perspectives and collaborations that bring both into focus. Jenny Odell's *How to Do Nothing* illustrates the ways the widespread focus on 24-7 work cycles and productivity in North America reduces our capacity to notice the world around us and, importantly, to change it. And James Wood's *Serious Noticing* brings these questions to bear on literary studies and illustrates how literature can sharpen our ability to notice, to understand different temporalities, and to appreciate the art (and politics) of perspective.

Part of noticing is also making explicit the things we do not tend to notice. Some of these things—the privilege some students have and others do not, systemic inequalities, the default toward linear temporalities, the invisibility of the infrastructures in which we live that uphold and promote carbon emissions, and so on—are reinforced by our reading. Indeed, literature often contributes to the shaping and naturalization of the infrastructures we inhabit. For middle-class North Americans, Raine writes, drawing on Stephanie LeMenager, these infrastructures often include "expectations of individual freedom, security, and prosperity made possible by colonialism, extractive capitalism, and unlimited burning of fossil fuels" (72). But literature can also train us to be sensitive readers of infrastructural assumptions and can give students the language and tools for thinking otherwise about many things long assumed unchangeable.

These comments bring me back to the contribution the humanities can make to climate change thinking and action. The humanities help students consider how culture shapes what and how they see, including the course they take with me. Gadsby closes *Nanette* by stressing the ways that stories connect us. While I try not to diminish the importance of Gadsby's focus on gender identity, I also try to extend her points about stories and the connections they enable and promote to topics like climate change that will also benefit from being more legible, relatable, and actionable in the classroom.[4]

Pedagogical Approaches

In the remainder of this essay I will focus on three pedagogical approaches—expertise, keyword work groups, and a one-day student-curated syllabus—that reflect and reinforce the foci of my course outlined above.

In any class, it's important to bring to the fore and empower what students already know. But in a course on climate change, in which students are already inclined to feel disempowered, if not defeated, this goal is especially important. At the beginning of class, I invite students to identify their area or areas of expertise. The point is to cast a wide net and not necessarily to articulate skills linked to climate change. I give them a few minutes to think about it and then we discuss their responses. One student is an expert at identifying trees, another is a skydiver, another plays the violin, another walks dogs, another jokes that she is good at packing a lot into a small suitcase, and another jokes that he's an expert at doing nothing. We have a general discussion together of these skills and the broader skills to which they may relate: memorization, endurance, mapping, practice, patience, planning, resting, and so on. And over the course of the semester we build on these areas of expertise and further develop them. I can turn to the student who knows about trees to comment on a passage in a novel that describes a forest, for example. I can return to the student who is an expert on doing nothing and, with his help and the help of others, deepen and expand on what both doing and nothing mean. I can turn to the student who is good at packing a lot of things into a small space to comment on an especially dense poem that also packs a lot into a small space. The students soon know each other's skills and add additional skills of their own to our discussion roster.

I try to illustrate how forms of expertise seemingly remote from climate change can be relevant to it. I also try to illustrate how each student has something to offer to the discussion. We are all experts in something, and when we share our expertise, we can learn from one another. I've found that when we move away from a single source of expertise—me, as the professor—and instead distribute expertise through the classroom, much more lively and productive discussions follow. Our treatment of the material is enriched by being refracted through the often surprising skills and expertise of the students, and this first-day assignment yields rewards that are felt throughout the semester.

Each week of my class is focalized through one keyword. The concept of keywords comes from Raymond Williams's influential book *Keywords* and Williams's larger project to illustrate how, as Jonathan Arac and Holly Yanacek put it, words are "repurposed" to capture "tumultuous changes, to think about them, and to act on them" (121). In this particularly tumultuous period of climate crisis, keywords can work as especially potent illustrations of how language carries meanings that adapt and shift historically and do particular types of work. The keywords, in short, both focus our thinking for a given day and illustrate the role of language in shaping both thought and action. Words can be dense nodes in which history is compacted and can also change through use and context.

We recall our discussion of noticing and think about how, when we have learned a new word, we often begin to see it everywhere. Robert

MacFarlane provides a striking example of how dictionaries respond to changing social conditions in his account of revisions made to the *Oxford Junior Dictionary* in 2007 and 2012. Many of the words that were removed (such as *acorn, ash, beach, beech, lark, mistletoe, pasture,* and *willow*) are associated with the natural environment, and many of the words that were introduced (*blog, broadband, chatroom, MP3 player,* and *voicemail*) are associated with new technologies (3). Are we less likely to see larks and willows if we don't have words for them? Do chatrooms become more important to our thinking than pastures when we only have a word for the former? As a class, we discuss how literary works—works that use words—shape and form the world in which we live, contributing to its possibilities and helping us orient ourselves within it.

In keeping with my goal to discuss words that reflect changing social issues, the course's keywords change from year to year. Some words that have remained constant since I developed the course three years ago are *nature, protest, migration,* and *fire*. But this year I also replaced some words I had used in previous years with, for example, *pandemic, pause, care,* and *normal*.

About halfway through each semester I ask students to select a short item, in any medium, that addresses climate change, in their view, effectively. I also ask them to write a one-page rationale for their choice and to situate it in the context of at least one item from the course syllabus. I then make these selections and rationales available to all the students.[5] I introduce this assignment early on so that students can be on the lookout for appropriate items. I'm always wonderfully surprised by the range and variety of their selections, from poems, short stories, and *YouTube* videos through songs, dance, visual art, and advertisements. This class offers an angle on our topic, drawing on the students' expertise and interests, that I could not possibly have formulated on my own. It also offers a demonstration of what we can accomplish through collaboration.

At the last class meeting, each student introduces their piece. The students are excited both by their own selections and by the selections of their peers. They become cocreators of the course and frequently identify this assignment as their favorite in course evaluations. For my part, I am struck by how what we teach is most vibrant and alive when it is not set apart from students' lives but rather springs from them and is in conversation with them.

By the end of the semester, my students typically feel less skeptical about the role of the humanities in addressing the climate crisis. They have illustrated, through their own expertise, discussions, attention to language, noticing, time, and angling, the vital role that the humanities and literature play in our comprehension of the climate crisis. Importantly, through our conversations and debates and, again, drawing on their own expertise and how it is amplified in dialogue with others, they come to appreciate that bringing as many voices and perspectives as possible (from the humanities as well as from science, policy, engineering, and business, among other fields) to bear on the climate crisis is one of our best hopes for forging an adequate response. And they experience how learning about and confronting the climate crisis is one of the best antidotes to the sense of disempowerment and anxiety climate change can otherwise produce. They also realize that, without discounting the gravity of our shared precarity, laughter is possible along the way.

Notes

Course syllabi are always collaborative endeavors, and I am indebted to many people in the design of the course described in this essay. I'd like to thank Anna Henchman, Caroline Levine, Benjamin Morgan, Anne Raine, and Stephen Siperstein for sharing syllabi and ideas. I'm also indebted to the article by Frances O'Gorman and colleagues on teaching the environmental humanities. And I'd like to thank the students who have taken this course and provided such wonderful input along the way. Special thanks to Rayan Eid and Skylar Chambers, in particular, for their stellar insights into how the course could be improved.

1. In a related vein, Anne Raine notes that ecocritical literary studies "trains students . . . to attend not only to what's explicitly stated but also to the feel of things, or what we might call the 'body language' of the text" (71).

2. To my mind, this latter point also means rethinking the traditional classroom, but that topic is beyond the scope of this short essay. While I do not teach climate change in detail, I do try to give students the resources to understand the science of climate change and the policy proposals that are most often discussed.

3. There is a huge amount of critical material on climate temporalities. Different Indigenous approaches to time are especially illuminating here (see Estes; Kimmerer; Whyte).

4. I also think it is important to articulate goals that are commensurate with the difficulty and urgency of the climate crisis. For example, Raine's learning goals for her class on climate change invite students "to consider how [internalized] norms participate in the capitalist, colonialist, and human-exceptionalist systems that produced the climate crisis; to practice expanding our attention from individual to systemic and more-than-human scales; . . . to cultivate compassion for and collaboration with diverse others, human and nonhuman; [and] . . . to experience ourselves not as isolated individuals but as co-emergent social-material actors with the capacity to contribute to systemic change" (74).

5. My class is composed of twelve to sixteen students; if it were larger, I would divide them into groups for this assignment.

Works Cited

Arac, Jonathan, and Holly Yanacek. "Keywords, Structures of Feeling, and the Novel." *Novel: A Forum on Fiction*, vol. 54, no. 1, 2021, pp. 221–25.

Estes, Nick. *Our History Is the Future: Standing Rock versus the Dakota Access Pipeline, and the Long Tradition of Indigenous Resistance.* Verso, 2019.

Kimmerer, Robin Wall. *Braiding Sweetgrass: Indigenous Wisdom, Scientific Knowledge, and the Teachings of Plants.* Milkweed Editions, 2013.

Macfarlane, Robert. *Landmarks.* Penguin, 2015.

Nanette. Written and performed by Hannah Gadsby, directed by Madeleine Parry and John Olb, Netflix, 2018.

Odell, Jenny. *How to Do Nothing: Resisting the Attention Economy.* Penguin, 2019.

O'Gorman, Frances, et al. "Teaching the Environmental Humanities: International Perspectives and Practices." *Environmental Humanities*, vol. 11, no. 2, 2019, pp. 427–60.

Raine, Anne. "Literary Reading, Mindfulness, and Climate Justice: An Experiment in Contemplative Ecocritical Pedagogy." *Contemplative Practices and Anti-Oppressive Pedagogies for Higher Education: Bridging the Disciplines,* edited by Greta Gaard and Bengü Ergüner-Tekinalp, Routledge, 2022, pp. 71–89.

Tsing, Anna Lowenhaupt. *The Mushroom at the End of the World: On the Possibility of Life in Capitalist Ruins.* Princeton UP, 2017.

Whyte, Kyle. "Indigenous Climate Change Studies: Indigenizing Futures, Decolonizing the Anthropocene." *English Language Notes*, vol. 55, nos. 1–2, 2017, pp. 153–62.

Williams, Raymond. *Keywords: A Vocabulary of Culture and Society.* Oxford UP, 1976.

Wood, James. *Serious Noticing: Selected Essays, 1997–2019.* Farrar, Straus and Giroux, 2019.

Tobias Menely

Possible Futures in a Warming World: Teaching Climate Models and Other Climate Fictions

Inverting *anachronism*, Srinivas Aravamudan came up with the term *cata-chronism* to capture the experience of living in a present defined by its relation to a future that is like the past in that it is already determined. Catachronism, he writes, "re-characterizes the past and the present in terms of a future proclaimed as determinate" (8). Actions that occurred in the past or are occurring now impose powerful constraints on future possibility, limiting options for individuals and societies to come. Cata-chronism recognizes the belatedness of our condition, the impossibility of fixing now what we might have fixed in the past. What was once con-tingent (say, the outcome of the 2000 US presidential election) has be-come necessity (two decades of growing emissions). Coal-fired power plants are being built today with an expected lifespan of forty years. As we enter the third decade of the twenty-first century and begin to witness the cat-astrophic implications of anthropogenic warming, global carbon emissions continue to grow each year. The carbon dioxide emitted today—and yes-terday, and tomorrow—will remain in the atmosphere for centuries, as will the heat captured in the oceans. Even if emissions are finally reduced, plan-etary warming and sea-level rise will continue for hundreds of years be-fore the global climate system reaches a new equilibrium. We know that the twenty-first century will bring a cascade of crises, disasters, and extinctions of a sort and at a scale unlike those human societies have faced heretofore.

A course on the literature of climate change, which for many university students also serves as an introductory course on the climate crisis, must confront this forbidding sense of necessity, of a future already written, which can easily fall into the pattern of preexisting apocalyptic imaginaries. Climate catachronism, Aravamudan observes, appears to express a "theological grasp of time, whereby anticipation, belief, and application on the present are integrated as inexorably leading to a known and inevitable outcome" (8). Climate fiction—an emergent genre, though one with deep roots in science fiction—can offer a secular alternative to a theological understanding of futurity, helping us imagine what it will mean to survive, to make do, to adapt, and to begin the work of repair in the difficult and damaged world to come. Climate fiction can reveal the textures of contingency, "that which is possible otherwise" (Wellbery 237), within a future that is (already) determined by anthropogenic climate change.

In teaching courses on climate change and literature, I find it useful to supplement climate fiction with another narrative form that represents potential planetary futures: climate models, and specifically the Representative Concentration Pathways (RCPs). The RCPs were developed to model a spectrum of potential Earth trajectories, showing how policy choices bring about different climate futures. They were first prominently featured in the fifth assessment report (AR5) released by the Intergovernmental Panel on Climate Change (IPCC) in 2014. Each of the four scenarios included in the report models a future for the Earth system defined by a single distinguishing variable: the amount of extra radiative forcing, measured in watts per square meter, caused by anthropogenic greenhouse gas emissions and changes in land use between the present and the year 2100. These scenarios—RCP 2.6, 4.5, 6, and 8.5—concentrate attention on the role played by greenhouse gases as the earth's thermostat. The pathways reflect the basic axiom of Earth system science: that our planet is an interconnected but also highly variegated system, homeostatic but also susceptible to significant nonlinear phase changes. Since AR5, three additional pathways (RCP 1.9, 3.4, and 7) have been introduced, as has a related model, the Shared Socioeconomic Pathways (SSPs). These models of potential futures have played an important role in global climate change governance (such as it is) over the past decade. The pathways allow the IPCC to convey the implications of political choices about greenhouse gas emissions, decarbonization, land use priorities, and economic growth to the public and to policymakers. As Julia Nordblad observes, the pathways, by vividly and precisely representing "alternative futures," provide "starting points for thinking politically"—for "imagining, creating, planning, or deliberating" (338, 330).

I begin my courses on climate change fiction with the RCPs because they reduce a complex system of interconnections between the Earth system and social systems to a shared nexus, allowing class members to formulate more precise questions about interconnections, inputs and outputs, and causes and consequences. We can ask, for example, how the pace of energy transition, or demographic changes, or feedback loops within the climate system, might place the earth on a given trajectory in the twenty-first century. We can then step back further and ask how political, economic, and technological variables inform the pace of decarbonization. We can move from society to the Earth system and ask how a given amount of warming will determine new disease vectors, agricultural productivity, the rate of sea-level rise, the distribution of extreme heat, desertification, and the likelihood of weather disasters. And then we can speculate about how human societies might respond—politically, demographically, and economically—to such transitions in the Earth system.

By working with the RCPs, we can also explore questions about narrative meaning-making—a concept familiar to my English students—in an unfamiliar way. In the first week of a seminar on climate change fiction, my undergraduates at the University of California, Davis, readily identify the pathways as narrative: a form for representing change over time. They characterize the unique narratological affordance of the RCPs, the economy with which these models represent multiple potential planetary futures. In the forking paths we see both contingency and necessity, the way in which any given planetary present is a result of and will itself become a determinate past at the same time that it remains open to multiple possible futures. I ask students how fictional narratives are and are not able to approximate the narrative work accomplished by the RCPs. Their answers are revelatory. One proposes that the principle of narrative conflict implies that events might unfold in a variety of ways. Another talks about how point of view can unsettle a reader's sense of a single clear line of causality. Focalization might function in a way similar to the determining variable of the RCP, providing a simplifying center point around which causal threads multiply. We discuss how characters (not unlike ourselves) define their choices by imagining future outcomes, their motives and decisions oriented by fictive futures, the stories they tell. A student brings up Catherine Gallagher's concept of "fictionality": the role realist but fictional stories played for eighteenth- and nineteenth-century readers in supporting "cognitive provisionality" (347) and the imaginative work of testing out life choices in a complex and rapidly changing society. Gallagher teaches us to understand "fictionality" as implying not unreality but potentiality; in this sense, we can see that the RCPs are also fictions. Students bring up jump cuts, counterfactuals, and the crucial role of surprise within narrative—a sudden collision of one system with another that bears some similarity to the nonlinearity of tipping points in the Earth system. We discuss examples of narrative forms—literary and otherwise—that stage multiple time lines: Jorge Luis Borges's extraordinary short story "A Garden of Forking Paths," the famous "Remedial Chaos Theory" episode of the television show *Community*, time loop films such as *Groundhog Day* and *Palm Springs*, the Choose Your Own Adventure series of children's books, board games, and computer games (which allow one to reset to initial conditions and play out a sequence differently).

During the term, we read a number of critical texts that help us expand our narratological lexicon, providing us with concepts for defining the specificity of climate change fiction and its relation to climate models. In *The Great Derangement*, Amitav Ghosh characterizes the "uniformitarian" conventions that have kept "literary" fiction from internalizing climate change: realism's rules of continuity, regularity, and plausibility and its fixation on human agency defined against a passive setting (35). Ghosh also invites us to consider why genre matters, not least in the difference between the bourgeois realism he associates with literariness and science fiction, which often does feature dynamic and uncanny worlds and extended timescales. In "Climate Change and Un-Narratability," Hannes Bergthaller argues that not just literary fiction but all imaginative narrative falls short: "climate, as a complex system with emergent properties, resists narrativization" (9). Unlike novels, models are designed to represent systemic complexity, feedback, and nonlinearity. Bergthaller helps us define precisely what models can accomplish that fiction cannot and what even models fail to represent. Returning us to the distinct affordances of fiction, Stephanie LeMenager's "Climate Change and the Struggle for Genre" shows how narrative fiction fills in the immersive and subjective features of climate futurity, the "everyday . . . lived time of the Anthropocene" that models leave out (225). LeMenager's definition of *genre* as a "pattern of expectation" turns our attention to the powerful work of narrative closure as well as the fascinating opportunity today to watch a new genre developing in real time in response to the pressures of the climate emergency. Christophe Bode, inspired by a chance encounter with climate models, theorizes a type of narrative that can not only *"thematize* openness, indeterminacy, virtuality, and the idea that every 'now' contains a multitude of possible continuations" but also find means of *"staging* the fact that the future is a space of yet unrealized potentiality" such that a reader can *"experience* that 'what happens next' may well depend upon us, upon our decisions, our actions, our values and motivations" (1). Whereas most narratives (even novels set in the future) organize events retrospectively, creating meaning from what has happened, future narratives present "nodes" where pathways into the future are selected. Bode views future narratives as an unusual narrative form (exemplified by the film *Run Lola Run*), but he provides us with terms to more precisely define the staging of "openness" or "contingency" within climate fictions and climate models.

The RCPs generate productive discussions about narrative form while also supplying a knowledge framework that deepens our ability to read and interpret climate fiction, especially when we turn our attention to a novel's depiction of setting (and especially when we observe setting shape plot). With the RCPs in mind, we discuss a novel's climate realism, its plausibility as a depiction of Earth's future, and even its implied trajectory. I always begin climate fiction courses with Octavia E. Butler's *Parable of the Sower*, set in California starting in 2024. We consider Butler's depiction of sea-level rise, drought, and wildfires, including a harrowing wildfire on what appears to be Highway 16, along Cache Creek on the way to Clear Lake, which is just northwest of Davis. The novel is compelling and clear-eyed in representing the relation between climate change and social breakdown: infrastructural failure, immiseration, disaster capitalism, and displacement. While it is not possible to correlate the novel with a precise RCP, it is clear that *Parable* represents a high-emission pathway. Kim Stanley Robinson's *New York 2140* includes a retrospective account of the disintegration of ice sheets and sudden sea-level rise of fifty feet late in the twenty-first century. Students point out that the degree of sea-level rise in the novel would be unexpected even in the 8.5 RCP. Robinson's novel is an occasion to introduce the Extended Concentration Pathways, which model climate through the year 2300, and to discuss the challenges of modeling (and imagining) potential future at that distance. Alexis Wright's *The*

Swan Book integrates novelistic realism with Australian Aboriginal story-work, allowing us to examine the realist conventions used to establish setting in time and place as well as Indigenous alternatives—cyclic, spiral—to a linear conception of history. Other novels that I teach—such as Barbara Kingsolver's *Flight Behavior* and Amitav Ghosh's *Gun Island*—are set in the near present, allowing the class to look for cues attesting to the pace of decarbonization. Our culminating work of climate fiction is a piece of experimental "speculative fabulation" (Haraway 136), Donna Haraway's "Camille Stories," which narrates four hundred years of planetary futurity focalized around five generations of human-butterfly symbionts (134–68). Haraway establishes a compelling balance between a hard-eyed reckoning with the extinctions to come and a hopeful account of the ongoing work of remembrance, reparation, and rebuilding. Her highly metanarratological fable also helps students think about the choices they might make in creating their own climate fiction.

During the final week, students present their own collaboratively developed climate fictions. Early in the term, I place students in groups of three or four and ask them to work toward a final project in which they outline a story—perhaps a pitch to an editor or producer—of a given future, extending from now until 2200, defined by one hard fact: a given degree of radiative forcing in 2100. Each group develops a narrative that explains why the world ends up following a particular RCP and what the implications of that RCP are for collective world-making. They are responsible for thinking about vectors of causality that begin with anthropogenic forcing (economic, technological, and political factors that lead to patterns of emissions and land use) and play out in geophysical consequences (sea-level rise, heat, storms, and ocean circulation) that generate demographic, economic, geopolitical, political, and cultural responses. They must figure out how to narrate these complex and multidirectional vectors of causality—on diverse scales, from individual experience to global totality—while also telling a compelling story. Finding this balance, in turn, allows them to distinguish (and so relate) the realist aspects of their narrative, its specific claim to plausibility, from its fictive components—namely, the choices they have made to emphasize certain points of view, certain identities, certain outcomes, and certain forms of human agency.

The presentations are, without fail, fascinating and inspiring, even as students reckon with the difficult and diminished future. Groups come up with an impressive range of narrative mediums to represent their scenario:

future NPR broadcasts, board games ("Climatopoly") and computer simulations, political manifestos, found diaries, multigenerational novels. We compare how groups with the same RCP differently solve the problem of accounting for their Earth system trajectory. The stories they tell are global in scope, often featuring widespread dislocation and migration, though they usually center on California. Some get deep into the specifics of decarbonization, describing the emergence of taxes on gas-powered vehicles, changing dietary and consumption patterns, and shifting land-use priorities in the Central Valley. The future scenarios include many of the disasters that have been present throughout human history: wars, pandemics, dearth, and mass death. They also include many surprising turns, featuring nonlinearity in social systems, planetary systems, and the socioecological interface. Like the most optimistic pathways, some rely on a true deus ex machina: scalable negative emissions technology or successful geoengineering initiatives. There is lots of hard thinking about collective action and political change, including the role of ecotage, activism (a "Mothers against Climate Change" movement), and the emergence of new forms of global sovereignty. But students also find ways to tell the stories of relatable characters navigating difficult change. This culminating assignment offers them a chance to collectively imagine the future in a way that acknowledges necessity while making room for contingency.

Works Cited

Aravamudan, Srinivas. "The Catachronism of Climate Change." *Diacritics*, vol. 41, no. 3, 2013, pp. 6–30.

Bergthaller, Hannes. "Climate Change and Un-Narratability." *Metaphora*, vol. 2, 2018, pp. 1–12.

Bode, Christophe. "The Theory and Poetics of Future Narratives." *Future Narratives*, edited by Bode, De Gruyter, 1993, pp. 1–108.

Butler, Octavia E. *Parable of the Sower*. Four Walls Eight Windows, 1993.

Gallagher, Catherine. "The Rise of Fictionality." *The Novel*, edited by Franco Moretti, vol. 1, Princeton UP, 2006, pp. 336–63.

Ghosh, Amitav. *The Great Derangement: Climate Change and the Unthinkable*. U of Chicago P, 2016.

———. *Gun Island*. Farrar, Straus and Giroux, 2019.

Haraway, Donna. *Staying with the Trouble: Making Kin in the Chthulucene*. Duke UP, 2016.

Intergovernmental Panel on Climate Change. *Climate Change 2014: Synthesis Report: Contribution of Working Groups I, II and III to the Fifth Assessment Report of the Intergovernmental Panel on Climate Change*. Edited by the Core Writing Team et al., IPCC, 2014.

Kingsolver, Barbara. *Flight Behavior*. HarperCollins Publishers, 2012.
LeMenager, Stephanie. "Climate Change and the Struggle for Genre." *Anthropocene Reading: Literary History in Geologic Times*, edited by Tobias Menely and Jesse Oak Taylor, Pennsylvania State UP, 2017, pp. 220–38.
Nordblad, Julia. "On the Difference between Anthropocene and Climate Change Temporalities." *Critical Inquiry*, vol. 47, winter 2021, pp. 328–48.
Robinson, Kim Stanley. *New York 2140*. Orbit, 2017.
Wellbery, David. "Contingency." *Neverending Stories: Toward a Critical Narratology*, edited by Ann Fehn et al., Princeton UP, 1992, pp. 237–57.
Wright, Alexis. *The Swan Book*. Atria Books, 2016.

Part VI

Hopefulness and Beyond

Kathryn Prince

Finding Hope in Climate Literature: Solastalgia, Twilight Knowing, and Unintended Consequences

The "hope gap" is a recognized factor inhibiting action that could address the climate emergency. Ironically, the work we do to educate students about climate change can contribute directly to the problem: recent studies have demonstrated a link between an increased awareness of this issue and feelings of powerlessness, hopelessness, and depression that can result in climate passivity, paralysis, or even denial (see Marlon et al.). Naming these and other emotional responses is an important part of how I approach the literature of climate change. As a scholar of emotions, my research often focuses on how texts practice emotions: how the words on the page name, communicate, mobilize, and regulate emotions, to use Monique Scheer's influential terminology. I often track the ways in which texts practice emotions and create "emotional communities" when they meet readers, sometimes across disparate times and places. While my work typically focuses on recuperating the emotions of the past in texts written long ago and thinking about the emotional communities that coalesce around them in subsequent periods, the urgent need to address the hope gap suggested to me that there might be value in seeking all this for my students within a different corpus: recent novels set in the future. My expectation, drawing on theories of how emotions circulate, was that novels might move readers to action through the mobilization of emotions and through the creation of emotional communities that connect not only writer and reader but also the present day and an imagined future in which different emotions are possible: a space William Reddy would describe as an "emotional refuge" (136). Against the hopeless narrative of an insoluble problem and an impossible future, I set these examples of life on the other side of the solution.

If it is no longer true that climate change is absent from literature, as Amitav Ghosh argued in 2016, we are still learning how to teach this growing body of work in ways that might help our students navigate a rapidly changing world. The climate emergency feels more urgent today than it did even when I taught this course in 2020, and so do our pedagogical and ethical responsibilities to the students for whom hope may be wearing thin.

The course that I shaped around these preoccupations was called Apocalypse, Now What? It began in February 2020, at the University of Western Australia, under a cloud of smoke from an especially devastating bushfire season. It shifted online within days in response to reports of a new coronavirus spreading worldwide. I had based my syllabus on the premise that the time of consciousness-raising was over, and these twin disasters served to underscore the point. I wanted students to imagine their way through and beyond the upheavals of climate change in order to find hope for the journey ahead—a journey that seemed, in that fraught moment, to be on the verge of beginning. The novels I selected to serve this purpose are not the most obvious examples of climate fiction, or even climate fiction at all, strictly speaking, but they share an emotional register that I associate with climate change and an interest in depicting life persisting after a major disaster on Earth.

This was an honors course in an English department, so it was possible to assign a complete novel each week, with an exception made for Neal Stephenson's sprawling *Seveneves*, which splits neatly into two parts, the second set five thousand years after the apocalyptic events depicted in the first. We began with Emily St. John Mandel's *Station Eleven*, which turned out to be more topical than I could have anticipated, its inciting incident a flu pandemic that eliminates some ninety-nine percent of the human population, reducing civilization to a few scattered outposts. While Mandel's novel as a whole is set in the aftermath of the pandemic, in the prepandemic flashbacks a key element is a graphic novel written and illustrated by one of the characters that links *Station Eleven*'s pervasive nostalgia for the prepandemic world to the emotion that Glenn Albrecht has named solastalgia, a longing for the lost landscapes of a home that has been transformed by climate change (27). In the graphic novel, Dr. Eleven's longing for his own ruined planet contrasts with the sparsely populated, bereft Earth inhabited by the characters in the novel proper. The beauty of Earth is not yet lost to the pandemic's survivors and may never be lost: Mandel's characters continue to recognize the beauty in their postpandemic world even as they mourn the lost trappings of civilization. In the real-life pandemic, news stories of biodiversity regaining lost ground as people retreated into lockdown underscored the potential, in the novel, of an Earth with a reduced human footprint.

Station Eleven is not "about" climate change, but it is a novel that deals profoundly with the climate change emotion of solastalgia, which it beautifully communicates and mobilizes, in Scheer's terms. Solastalgia forms an emotional community within the novel, and it exerts a dangerous appeal by anchoring the characters to their lost past, ultimately bringing the apparently disparate narrative threads together in a conclusion that shows the surviving characters looking straight ahead into the future, not longing for what has been lost. In this way, the novel offers the possibility of an emotional community that includes its readers, one that regulates backward-looking solastalgia in order to mobilize forward-looking hope.

Overall, my objectives for the course were to understand how hope for the future, even in the face of disaster and loss, is mobilized in these novels. Not all the students were able to believe in the emotional communities formed by *Station Eleven*'s survivors, housed in an abandoned airport where a lounge has become a museum commemorating lost technologies and people; the Traveling Symphony of musicians and actors keep alive what was best of the civilization now ending because they believe it is worth preserving. The possibility of a more dystopian outcome is often suggested in the novel, which even seems to allude to Cormac McCarthy's postapocalyptic novel *The Road* in some passages. Some students offered *The Road* as a more realistic counterpart to *Station Eleven*, so I recruited Rebecca Solnit's *Paradise Built in Hell* to bolster the novel's case that disaster can lead to positive emotions centered on the ideas of mutual aid and community. Even reading novels featuring humans who evolve epigenetically in response to their experiences (Stephenson's *Seveneves*), hybrid creatures like pigoons and rakunks genetically engineered in terrifying corporate compounds (Margaret Atwood's *Oryx and Crake*), and a Galactic Commons comprising diverse spacefaring species like the reptilian Aandrisks and the sluglike Harmagians (Becky Chambers's *Record of a*

Spaceborn Few), students found the postcapitalist barter economies in the Mandel and Chambers novels, and in Kim Stanley Robinson's *New York 2140*, the most implausible. They were better able to imagine postapocalyptic alternatives to capitalism in the context of Waubgeshig Rice's near-future novel *Moon of the Crusted Snow*, in which an unnamed disaster cuts off an Indigenous community already dealing with unreliable supplies of food and electricity from "down south." The idea of living in harmony with nature, of taking nothing from the land except what is necessary for survival, always with reverence and respect, is a profoundly ecological and ethical position consistent with climate goals (see United Nations). By mobilizing positive emotions around the precedent of traditional Indigenous knowledge, Rice's novel regulates the kind of skeptical response the other novels seem to have elicited and offers the possibility of an emotional community in the future built around Indigenous cultural practices. After all, as a character in Rice's novel points out, "[A]pocalypse. We've had that over and over. But we always survived. We're still here. And we'll still be here, even if the power and the radios don't come back on and we never see any white people ever again" (130).

Students had the option to engage with these novels critically or creatively, and I found that introducing secondary reading to scaffold the fictions theoretically and factually was no more effective than inviting students to extend them imaginatively. Students who took the creative option by writing a bonus chapter for one of the novels were able to engage more closely with their narrative strategies and to observe how these communicate, mobilize, and regulate emotions: they emulated the recursive narrative structure of *Station Eleven*, the lacunae of that novel and *Seveneves*, and the narratorial perspectives of *Oryx and Crake* and *New York 2140* to create sophisticated emotional effects. Further, students offered well-informed critical perspectives about Indigenous knowledge, responding to *Moon of the Crusted Snow* and Claire G. Coleman's *Terra Nullius*; economic theories, responding to *New York 2140*; and reproductive science, responding to *Seveneves*, that incorporated acute awareness of emotional practices and emotional communities.

One novel that came out too late to be included in the course but that informed our discussions to a significant degree is Jenny Offill's beautiful climate fiction novel, *Weather*. Beneath the narrator's retelling of other people's anecdotes—momentous, monstrous, and mundane—is an occasional glimmer of climate panic, most discernible in the stories Lizzie tells about herself. "Environmentalists are so dreary," she complains (51), but she becomes a veritable climate Cassandra, a "crazy doomer" (88), when drinking or depressed, obsessed with disaster psychology and "prepper things" (146). She is living in a state that Stanley Cohen describes as "the twilight between knowing and not-knowing" (80) or "twilight knowledge" (81) in his influential work on the sociology of atrocities, *States of Denial.* This knowing-not-knowing is coming to an end in the novel for Sylvia, a process she describes to Lizzie as "losing heart" (Offill 83), which manifests in her struggle to find the "obligatory note of hope" (65, 140) that once came easily to her. Eventually, Sylvia resigns from her leading role at an environmentalist foundation because "there's no hope anymore, only witnesses" (132). In retrospect, the anecdotes Lizzie relays as the narrator can be read as a way of bearing witness, of moving on from twilight knowing to full recognition and even, though the novel remains ambiguous about this, to some kind of action. Lizzie's husband, Ben, certainly takes this leap, at least emotionally, in the conclusion: he has done the same calculations she has about their son's possible future, "and now there's a quote from Epictetus pinned above his desk," Lizzie observes: "*You are not some disinterested bystander / Exert yourself*" (194). These characters' trajectories in the face of intolerable knowledge about climate change profoundly influenced the course.

What keeps students in twilight knowing may not always be a lack of awareness, solved through the dissemination of factual information, but rather a lack of hope, which locks them in the state of passivity Cohen observed and is exacerbated by climate-awareness activities that hammer home the same desperate message of imminent disaster. In order to find hope, students need to be able to imagine a life after fossil fuels, through and past the sixth mass extinction, that may be less comfortable than what they have known: a realistic sort of hope, not the hope in escapist fiction, the toxic illusion distracting us from a real chance to save ourselves. There is nothing wrong with reading for pleasure, and during the pandemic it has been one of the few safe havens into which students have been able to retreat, but my goal was something different.

Some students in my course, invigorated by hope, were dismayed by the gap between our conflicted present and the hopeful (or semihopeful, or potentially hopeful) future these novels depict. They wanted to engage with materials that could serve as the springboard for action, not in some fictional future but now, in the present, when it matters most. This is a salutary response, according to Saul Alinsky's influential work on activism, *Rules for Radicals.* Hope without action can be indiscernible from wishful thinking, "the simple hope that everything will work out some-how, some way" (xvi). Alinsky favors an activist hope anchored in "a future with a purpose" and "a will to fight for a better world" (21). With that idea of activist hope in mind, the next time I teach a course like this I will switch Kim Stanley Robinson's *New York 2140* with another, his vision-ary 2020 novel *The Ministry for the Future*, which imagines what the near future could be like if climate activists managed to implement some of the good ideas we already have that might help us create a future worth having.

Robinson's ideas around geoengineering, economic justice, and indi-vidual and collective action perfectly capture the tension between produc-tive and destructive hope that became a recurring concern in this course. In an interview broadcast on in July 2021, just as the North American West Coast was sweltering in a deadly "heat dome" eerily reminiscent of the devastating heat wave in India that opens *The Ministry for the Future*, Rob-inson explained that *Ministry* was intended to demolish what he now saw as a source of destructive hope, the idea that we might adapt to a chang-ing climate—the very idea that formed the theme of his earlier novel *New York 2140* ("Imagining Climate Futures"). As events from Oregon to Alaska added to the evidence for Robinson's position, he described his one re-maining utopian dream not as the citizen-led climate justice that *New York 2140* depicts but simply "to dodge a mass extinction event" through geoen-gineering. His terminology evokes Holly Jean Buck's *After Geoengineering: Climate Tragedy, Repair, and Restoration*, which, like Robinson, argues that dodging extinction will require a range of social and technological innovations, not one silver bullet solution. While Robinson's solutions are controversial within the loose emotional community around climate ac-tivism, it is increasingly clear that urgent action is required. Climate de-spair is a real option in the face of mounting evidence like this, and novels like Robinson's provide an antidote by mobilizing hope.

My own utopian hope is now not only to inoculate students against despair using hope, not only to move them from twilight knowing to en-gagement, but also to help them think through the dangerous hope in-herent in technological solutions to the climate crisis. When that heat dome settled over the West Coast, I wondered what might have ensued had that event occurred a year earlier. President Trump, a proponent of geoengi-neering who reportedly proposed disrupting hurricanes using nuclear bombs, could conceivably have gone rogue in just this way (see Stone). A tech billionaire could still do so at any moment using little more than pocket change and relatively simple technology to achieve a temporary "Pinatubo effect" (see Hawaiian Volcano Observatory). Robinson offers an optimistic vision of what might happen next, but Megan O'Keefe's Protectorate trilogy depicts a disastrous outcome that would be worth considering alongside Robinson's to imagine the emotions of regret that her novels mobilize.

The world has changed since I designed this course in early 2020. If I were to teach it again, I would include Robinson and the first book in O'Keefe's trilogy, *Velocity Weapon*, and I would make Offill official. I would hope to do this in a classroom full of not only English majors but also engineering and science students, working together to grapple with the issues these novels raise. A key plot point in O'Keefe's trilogy is that stories function as a data set through which empathy is learned, and I believe that sharing stories about unintended consequences, solastalgia, and twilight knowing across disciplines may be the most effective individual action I can take in the climate emergency. It would appear that I, too, am looking for hope.

Works Cited

Albrecht, Glenn. *Earth Emotions: New Words for a New World*. Cornell UP, 2019.

Alinsky, Saul. *Rules for Radicals*. Random House, 1971.

Atwood, Margaret. *Oryx and Crake*. McClelland and Stewart, 2003.

Buck, Holly Jean. *After Geoengineering: Climate Tragedy, Repair, and Restoration*. Verso, 2019.

Chambers, Becky. *Record of a Spaceborn Few*. HarperCollins Publishers, 2018. Wayfarers 3.

Cohen, Stanley. *States of Denial: Knowing about Atrocities and Suffering*. 2001. Blackwell, 2013.

Coleman, Claire G. *Terra Nullius*. Hachette, 2017.

Ghosh, Amitav. *The Great Derangement: Climate Change and the Unthinkable*. U of Chicago P, 2016.

Hawaiian Volcano Observatory. "Volcano Watch—The Pinatubo Effect: Can Geoengineering Mimic Volcanic Processes?" *USGS: Science for a Changing World*, United States Department of the Interior, 17 Feb. 2011, www.usgs .gov/center-news/volcano-watch-pinatubo-effect-can-geoengineering-mimic -volcanic-processes.

Mandel, Emily St. John. *Station Eleven*. HarperCollins Canada, 2014.

Marlon, Jennifer R., et al. "How Hope and Doubt Affect Climate Change Mobilization." *Frontiers in Communication*, vol. 4, 21 May 2019, https:// doi.org/10.3389/fcomm.2019.00020.

McCarthy, Cormac. *The Road*. Alfred A. Knopf, 2006.

Offill, Jenny. *Weather: A Novel*. Knopf Doubleday, 2020.

O'Keefe, Megan. *Velocity Weapon*. Orbit, 2019.

Reddy, William M. *The Navigation of Feeling: A Framework for the History of Emotions.* Cambridge UP, 2001.

Rice, Waubgeshig. *Moon of the Crusted Snow.* ECW Press, 2018.

Robinson, Kim Stanley. "Imagining Climate Futures with Kim Stanley Robinson." Interview by Azeem Azhar. *Harvard Business Review,* 7 July 2021, hbr.org/podcast/2021/07/imagining-climate-futures-with-kim-stanley-robinson.

———. *The Ministry for the Future.* Orbit, 2020.

———. *New York 2140.* Orbit, 2017.

Scheer, Monique. "Are Emotions a Kind of Practice (and Is That What Makes Them Have a History)? A Bourdieuian Approach to Understanding Emotion." *History and Theory,* vol. 51, no. 2, 2012, pp. 193–220.

Solnit, Rebecca. *A Paradise Built in Hell.* Penguin, 2009.

Stephenson, Neal. *Seveneves.* William Morrow, 2015.

Stone, Madeleine. "Beyond Nukes: How Scientists Dream of Killing Hurricanes." *National Geographic,* 27 Aug. 2019, www.nationalgeographic.com/environment/article/beyond-nukes-how-scientists-dream-of-killing-hurricanes.

United Nations, Department of Economic and Social Affairs: Indigenous Peoples. "Climate Change." www.un.org/development/desa/indigenouspeoples/climate-change.html. Accessed 27 July 2021.

Brianna R. Burke

Ruin, Rebellion, Remaking: Environmental Justice in the Literature of Climate Change

For nearly a decade, my environmental literature classes aimed to answer two sprawling and interconnected questions: how will climate change exacerbate preexisting ideologies of inequality, such as racism, classism, sexism, and speciesism, and how does the literature of climate change show which bodies—both human and other-than-human—are deserving of rights or protection in the midst of a quickly escalating ecological crisis? Each semester, my students and I approached these questions through exploring the effects of climate change alongside our texts and theories of the Anthropocene, environmental justice, and critical animal and animality studies. And each semester, students asked how I "coped" with what I knew and taught them, how I continued to work at the intersection of environmental injustice and climate change when it seemed as if both only became more deeply entrenched with each passing year. It was a good question, one that illuminated significant lacunae in my courses. Those of us who teach about environmental precarity, extinction, climate change, or social injustices regularly steep ourselves in the most devastating human struggles, which can indeed be overwhelming. As the years passed, the despair and cynicism my students brought into the classroom with them increased, becoming heavier; hopelessness saturated their responses to course material.

And then the pandemic hit.

The concatenated crises of climate change, globally entrenched and politically sanctioned inequality, and a global pandemic, mutually entangled and amplifying, were too enveloping. Some of my students fell sick; some also lost relatives or became primary caretakers, and all of them attended class virtually while in quarantine, their faces carefully neutral. In this new era of teaching, those of us who ask our students to critically examine and think deeply about endemic social, cultural, and environmental problems encountered additional challenges. As I watched students struggle to understand a new global health crisis, wrestle with difficult material, and simultaneously worry about their narrowing futures, I revised my class and explicitly began to teach how I "cope" as a scholar and teacher of what I call "cannibal economies" and bodily precarity exacerbated in the age of climate change. Therefore, this essay will discuss how I teach resilience, hope, and beauty, buoying students—and myself—so that we can engage in the hard work of understanding how cultural ideologies have created our planetary situation and, most importantly, how we can build a community of people working for change, imagine new and differently conducted human lives, and tell new stories.

I had been slowly altering my course over time to include more texts and assignments that required students to explore how change is created, but the pandemic caused me to fundamentally reexamine how I teach. Students are inundated with narratives of panic, fear, horror, and doom. Yes, I understand we are the verge of planetary collapse, and, like all scholars of climate change science and humanities, I know that public information about the acceleration of climate change is conservative. The greenhouse gases currently warming our planet were emitted decades ago; we haven't even begun to feel the greenhouse gas emissions of the past decade. I do not want to perpetuate denialism, to moderate these realities, or to create false promise. At the same time, it is irresponsible to contribute to narratives of fear and hopelessness in our classrooms. Students already know them; they are all too familiar with fear, and, like any of us, they can be paralyzed by it, frozen in place, when what we need is exactly the *opposite.*

I begin by talking about building tools of resilience from the very first class, telling students that we will be looking at how corporeal vulnerability is unequally distributed according to species and the cultural ideologies of race, class, and gender and that, because studying these topics requires intellectual and emotional heavy lifting, we will purposely engage in community building and embrace fully the joy of learning through art, literature, and culture. Literature inherently asks readers to enter into, to become; it can provide a safe harbor in which to examine issues that in other forms may be too threatening to face. In addition, as Lewis Hyde discusses in *The Gift*, literature is a gift from one person to another, given freely, without obligation. Art and literature exemplify the very best of our species, including the creative genius, the urge to understand, and the desire to communicate. When we get lost in the content, I bring students repeatedly back to this context, pointing out that each text is an example of yet another person working for change. Throughout the course, we purposefully seek examples of collective action, of communities embracing hope and fostering resilience. As Toi Derricotte wrote, "Joy is an act of resistance," and so, I tell students, we will cultivate joy and purpose, in and in company with one another.

My classes are inherently transdisciplinary, cross-listed with environmental studies, and so draw a wide range of students from every major at a large state institution of science and technology. Students from the arts, sciences, and humanities take my class, which is designed with flexible units to accommodate shifts in departmental representation from one semester to another. As a result, I do not teach all the texts in each unit described below but select from among them in a given semester. All the assignments are also flexible, allowing students to choose the options that best fit their majors, future professional goals, or personal interests. Such flexibility allows the multiple knowledges and interests in my classrooms to flourish and is the best part of teaching a truly interdisciplinary class; because students supply so much of the content, our collective understanding of the material is expansive. The purpose of the class remains stable and is relatively simple: to establish what, exactly, the environmental humanities contributes to understanding contemporary environmental issues and what role literature can play in accruing knowledge.

Unit 1: Impacts, Ideologies, and the Environmental Humanities

The first unit of my class is ambitious, exploring the sprawling effects of climate change and what the humanities add to environmental discourses while also initiating a semester-long conversation identifying the cultural ideologies that fuel the climate crisis. Texts revolve around the terms *climate change* and *Anthropocene* and include "The 'Anthropocene,'" by Paul Crutzen and Eugene F. Stoermer; "Thinking Like a Mountain," by Aldo Leopold (129–33); chapter 2 of Edward O. Wilson's *The Future of Life* (22–41); chapter 9 of Alan Weisman's *The World without Us* (112–28); Elizabeth Kolbert's essay "The Sixth Extinction?"; and Andrew Ross's essay "Climate Change." As they work their way through these texts, students choose from the following options for their first assignment:

> Explore an effect of climate change on local, state, national, or global ecologies.
>
> For a week, keep a record of everything you throw away or consume, and do a life cycle assessment. Learn about the impact of either production, consumption, or disposal and multiply impacts over a period of time of your choosing and population sample of your choosing.
>
> Conduct an energy audit for a week, recording all energy you consume from all sources and tracing one to three of those sources along the same lines as the assignment above.

While the students work on gathering data and producing their assignments, we watch *Wall-E* and read "The Great Silence," by Ted Chiang, ending the unit with class presentations.

Students reel from the information they learn in this unit, which is just a tiny segment of the scope of our sprawling planetary problem. Mindful of their grief and concern, I assign free writing, meditation, and what I call "Miraculous Earth Minutes," short videos that highlight the beauty, wonder, and weirdness of our planet or teach us about groups working on small ecological projects in different parts of the world. Students write about their climate change fears, concerns, or worries and about why particular environmental issues are important to them. As a class, we share our answers, and later I bring back into class common themes from their writings. For example, this year, for the first time, over seventy percent of my students noted that one of their "climate fears" revolved around having children, many expressing a fear that it is "irresponsible" to do so in this age. I agree with Ross's essay that we must be careful about the "neoliberalization of climate guilt," in which corporations and systems evade responsibility for the climate crisis by "outsourcing" their contributions "onto the conscience of individuals," resulting in what Ross calls "a new kind of moral tyranny" (40). However, climate change is anthropogenic, rooted in human behavior; this means that human behavior is relevant, even crucial. It also means that climate change is not inevitable or immutable. Human behavior can and often does change. And since analyses of culture, human behavior, values, and ideologies are precisely what the humanities bring to any environmental conversation, my students begin to understand this work as critical to our future on this planet.

Unit 2: Environmental Justice and Multispecies Interconnectivity

At the end of the first unit, we segue into environmental justice, extending our conversation about the Anthropocene and climate change. In the second unit, we explore how bodily precarity is socially and economically engineered. Texts include the Commission for Racial Justice's report *Toxic Wastes and Race in the United States*; "Animal," by Stacy Alaimo, and "Environmental Justice," by Giovanna Di Chiro, from *Keywords for Environmental Studies* (Adamson et al.); "Learning the Grammar of Animacy," by Robin Wall Kimmerer, from *Braiding Sweetgrass* (48–59); "On the Importance of a Date; or, Decolonizing the Anthropocene" by Heather Davis and Zoe Todd; *Sila*, by Chantal Bilodeau; chapter 1 of *The Truth about Stories*, by Thomas King (1–29); (sometimes) *Oryx and Crake*, by Margaret Atwood; and (sometimes) *Parable of the Sower*, by Octavia E. Butler. I offer students the following assignment options:

> A short paper on an essay of your choice from *Keywords for Environmental Studies*
>
> A short paper on the United Nations Declaration on the Rights of Indigenous Peoples
>
> A brief research paper on any aspect of the effects of climate change on the Biloxi-Chitimacha-Choctaw in Louisiana: coastal erosion, oil infrastructure in the region, their efforts to secure federal funds to relocate, their struggle for federal recognition, their plans to build a new village, or any other aspect of their dispossession
>
> A brief research paper on any aspect of the effects of climate change on the Inupiat village in Kivalina, Alaska: coastal erosion, their efforts to secure federal funds to relocate, their plans to build a new village, or any other aspect of their disposession

This unit focuses specifically on how the impacts of climate change are unequally distributed and examines struggles for climate justice. Indigenous nations were the first peoples to experience displacement and dispossession on this continent, and they continue to experience "environmental dis-possession," as Rob Nixon puts it in *Slow Violence and the Environmentalism of the Poor*, through resource extraction and climate change (5). In this unit, we also return to a conversation of multispecies kinship systems and how conjoined precarity works in the age of the sixth mass extinction, a discussion started with Ted Chiang's "The Great Silence." As Davis and Todd argue, the ideologies that engendered the Anthropocene are rooted in colony and empire; it is important to make it clear to students that the United States was built as, and continues to be, a settler colony. Kimmerer, King, and Alaimo add discussions of other-than-humans or interspecies connectivity to our conversations about human social injustices. Discussion of climate change or the Anthropocene tends to focus on human beings as both agents and victims, yet the writers in this unit point out that settler colonialism, in dominating a landscape and all its beings, imposed a hierarchy where humans are ascendant and all else rendered into resources. Meanwhile, species boundaries are porous, and humans and communities considered inhuman—animal—can become just another resource for consumption. This is what "conjoined ecological and human disposability" looks like, as Nixon calls it (4). These ideologies have long histories and continue to have echoing consequences in the world.

Unit 3: Remaking the World

Our class takes a deliberate turn in this unit, focusing on reckoning, collective action, and rebuilding. Texts include *Awake: A Dream from Standing Rock*; the website *Honor the Earth* (honorearth.org), particularly any media about the current and ongoing fight over the Enbridge 3 pipeline (just to the north of where I teach in Iowa); *The Back of the Turtle*, by Thomas King; and "Apocalypse: What Disasters Reveal," by Junot Díaz. Assignments include

> Game changers paper on individuals, groups, or companies working
> for environmental or cultural change
> Collective action paper on groups of people who have fought for, and
> instituted, positive environmental or cultural change in their areas
> Community action paper on changes our university, town, or state
> has instituted working for positive environmental or cultural change

The continuing focus on Indigenous peoples and issues in this unit resonates with my work in American Indian studies and with the resource rebellions occurring in the Midwest, where we live and learn. Just to the north of us, Indigenous peoples are trying to protect not only their own homes and cultures, along with the immense biodiversity around them, but also those of us who currently live literally downstream. King's book *The Back of the Turtle* is a breath of fresh air for our class as well; brilliant, funny, and charming, it argues that American society rewards psychopathy—the relentless and ruinous seeking of profit above all else—with wealth and prestige but that atonement for environmental damage is still possible. The novel calls for first reckoning with the damage wrought, then admitting one's part in the damage, and finally reinvesting through community building and ecological restoration, creating spaces for multispecies thriving. At the end of the novel, the work of remaking the world is not done, not finished—but then, creation never is; it is always unfolding.

Unit 4: Ruin, Rebellion, and Rebuilding

Finally, in the last unit, I turn toward Iowa, a landscape dominated by corporate agriculture. Texts include *The Hunger Games*, by Suzanne Collins; the introduction to *How the Other Half Dies*, by Susan George (3–27); the introduction to *Stuffed and Starved*, by Raj Patel (1–19); *Earth Democracy*, by Vandana Shiva; and the Dalai Lama's address memorializing Mahatma Gandhi and radical compassion. I have written at length about *The Hunger Games* elsewhere and have taught the novel for years. It remains surprisingly relevant. Its arguments that food is an environmental justice issue, that human beings depend on one another and on our environments to survive, that the subjugation of the poor through the consumptive desires of the wealthy is unjust, and that compassion is contagious—that it has, in fact, the power to incite rebellion—address the problems of our age as well as the ethos of our university and state, which purports to "feed the world." With *The Hunger Games*, I show students how the area where we live and learn, in the Midwest—the "flyover states"—is an extractive geography, and how we all live with the repercussions of corporate agriculture. Our water is polluted; our soil is depleted; our air is not clean. The pretty ethos of "feeding the world" hides the reality that we are no more immune to the ideologies of disposability or corporate greed than Katniss and the characters in *The Hunger Games*.

The ideologies we have been exploring from the start rebound, turn inward, and become cannibalistic. In the face of this reality, we come together and follow the examples set by the characters we have met, by each writer we have read.

Implementing Knowledge

In the midst of their journey south in Cormac McCarthy's *The Road*, the man and the boy "come upon the possessions of travelers abandoned in the road years ago. Boxes and bags. Everything melted and black" (190). They thread their way carefully through these abandoned symbols of flight, desperation, and capitalism, finally coming upon "the dead. Figures half mired in the blacktop, clutching themselves, mouths howling" (190). The boy asks, "Why didn't they leave the road?" And the man replies, "They couldn't. Everything was on fire" (191). This singular moment encapsulates the *The Road*: a parable about being trapped on *the* road already laid, one of increasing consumption, impact, and ruination. In the novel, the planet has died; there is no way to "leave" the road—the destination has already been reached. Yet McCarthy doesn't leave his readers on the road, in the world he has imagined; instead, he leaves them in this world, with a brook trout in their hands, gazing "on their backs" at the "vermiculate patterns that were maps of the world in its becoming. Maps and mazes. Of a thing which could not be put back" (287). I will not leave my students on the road, either—for we need to step off the road, chart new paths. Like the man in McCarthy's novel, those with strength and knowledge have a duty to shepherd the young and innocent. It is true that we are desperately short on "answers" or comprehensive solutions and that the true answers are both simple and complex: a radical reimagining of what it means to be human on this planet, a rewriting of cultural values on a vast scale.

Last year in class when talking about multispecies communities and extending rights or protections to the environment and other-than-human beings, one of my students said, "Sure, all of this sounds nice, but we aren't ready for that yet." I asked, "When you say 'we,' who do you mean?" She responded, "Humans." "All humans?" I asked. There was a collective nod of agreement on my computer screen. I urged the students to examine this grim assessment of humanity as unwilling to make collective change and to reimagine what the world could be. Most of them, I noted, had declared in class that they were ready, as individuals, to extend rights, to give up privileges, to work hard, to sacrifice to make a better, more equitable, more just and healthy world. If they were telling the truth in those moments, I proposed, what made them think that they were singular, either uniquely special or completely isolated and alone, in that readiness? "Aren't *you* part of the 'we'?" I asked.

At the end of the semester, in my concluding lecture, I ask students, What is education for, if not to illuminate the world and their relation as individuals to the whole? How will they reshape their lives to reflect what they have learned, what they now know to be true? After all, we have just spent a semester thinking about a ruin, rebellion, how to remake the world. At the end of *Rediscovery of North America*, Barry Lopez writes that we must "turn to each other and sense" that change "is possible" (53). Comments like this anonymous course evaluation allow me to hope that my students receive that message: "Her conviction in the importance of the work, that change is possible and dependent upon on our commitment, made what otherwise could have felt like a theoretical exercise in depressing futility and grief into an empowering masterclass in radical hope." I ask my students to look at each other, reflect on all the work we have done over the semester, and believe that change is not only "possible" but already underway.

Works Cited

Adamson, Joni, et al., editors. *Keywords for Environmental Studies.* New York UP, 2016.

Alaimo, Stacy. "Animal." Adamson et al., pp. 9–13.

Atwood, Margaret. *Oryx and Crake.* Anchor Books, 2004.

Awake: A Dream from Standing Rock. Directed by Myron Dewey et al., Digital Smoke Signals / International WOW Company, 2017.

Bilodeau, Chantal. *Sila.* Talonbooks, 2015.

Butler, Octavia E. *Parable of the Sower.* Seven Stories Press, 2017.

Chiang, Ted. "The Great Silence." *Exhalation: Stories,* by Chiang, Alfred A. Knopf, 2019, pp. 231–36.

Collins, Suzanne. *The Hunger Games.* Scholastic, 2010.

Commission for Racial Justice. *Toxic Wastes and Race in the United States: A National Report on the Racial and Socio-Economic Characteristics of Communities with Hazardous Waste Sites.* United Church of Christ, 1987.

Crutzen, Paul, and Eugene F. Stoermer. "The 'Anthropocene.'" *The Global Warming Reader: A Century of Writing about Climate Change,* edited by Bill McKibben, Penguin Books, 2021, pp. 69–72.

Dalai Lama. "Mahatma Gandhi." *Asiaweek,* www-cgi.cnn.com/ASIANOW/ asiaweek/98/0612/sr3.html. Accessed 25 Apr. 2012.

Davis, Heather, and Zoe Todd. "On the Importance of a Date, or Decolonizing the Anthropocene." *ACME: An International Journal for Critical Geographies,* vol. 16, no. 4, 2017, pp. 761–80.

Derricotte, Toi. "Joy Is an Act of Resistance." *Prairie Schooner,* vol. 82, no. 3, 2008, pp. 22–27.

Díaz, Junot. "Apocalypse: What Disasters Reveal." *The Boston Review*, 1 May 2011, www.bostonreview.net/articles/junot-diaz-apocalypse-haiti -earthquake/.

Di Chiro, Giovanna. "Environmental Justice." Adamson et al., pp. 100–05.

George, Susan. *How the Other Half Dies: The Real Reasons for World Hunger.* Penguin Books, 1985.

Hyde, Lewis. *The Gift: Creativity and the Artist in the Modern World.* Vintage Books, 2007.

Kimmerer, Robin Wall. *Braiding Sweetgrass: Indigenous Wisdom, Scientific Knowledge, and the Teachings of Plants.* Milkweed Editions, 2013.

King, Thomas. *The Back of the Turtle.* HarperCollins Publishers, 2014.

———. *The Truth about Stories: A Native Narrative.* House of Anansi Press, 2003.

Kolbert, Elizabeth. "The Sixth Extinction?" *The New Yorker*, 18 May 2009, www.newyorker.com/magazine/2009/05/25/the-sixth-extinction.

Leopold, Aldo. *A Sand County Almanac.* Ballantine Books, 1990.

Lopez, Barry. *The Rediscovery of North America.* Vintage Books, 1990.

McCarthy, Cormac. *The Road.* Alfred A. Knopf, 2006.

Nixon, Rob. *Slow Violence and the Environmentalism of the Poor.* Harvard UP, 2013.

Patel, Raj. *Stuffed and Starved: The Hidden Battle for the World Food Supply.* Melville House, 2007.

Ross, Andrew. "Climate Change." Adamson et al., pp. 37–41.

Shiva, Vandana. *Earth Democracy: Justice, Sustainability, and Peace.* North Atlantic Books, 2015.

Wall-E. Directed by Andrew Stanton, Walt Disney Animation Studios, 2008.

Weisman, Alan. *The World without Us.* St. Martin's Press, 2007.

Wilson, Edward O. *The Future of Life.* Alfred A. Knopf, 2002.

Ria Banerjee

Now What? Moving Past Climate Change Anxiety in an Interdisciplinary Community College Classroom

This essay discusses strategies I've used in an interdisciplinary, writing-intensive capstone course examining climate change through a humanistic, creative lens. I had some flexibility in choosing a topic and tailoring outcomes for the capstone, which I titled Disaster; Now What? Interdisciplinary Approaches to Climate Change. The still-popular rhetoric about avoiding or stopping climate change is defunct and unhelpful; instead, I designed this course to focus on ways to live better through impending climate crises.[1] My broader objective was to anticipate with students how planetary changes will intimately affect human life as we know it. During the course, we read widely in fiction as well as nonfiction from popular science writing and social sciences in order to make explicit how climate change is an intersectional, interdisciplinary problem that can be addressed through various academic lenses that students find most compelling. Acknowledging "disaster" head-on allowed us to move past defensive or apathetic reactions to climate science and focus on taking intentional actions in future. My students recalled New York City's uneven recovery from the devastation of 2012's Hurricane Sandy; connecting those past experiences to our course, we were able, at a discursive level at least, to anticipate the shape of our futures.

The framing and organization of this course grew from my observation that a hopeful/hopeless paradigm obscures the real work of learning to live with climate crises. Embracing "disaster" from the outset allowed me to direct student attention away from the ubiquitous rhetoric pushed on us by "greenwashed" commercials that offer visions of unchanged consumerism—and away from the obverse of these ideas, an equally conservative and passive hopelessness. In a recent interview that we read for class, the philosopher Donna Haraway makes a distinction between "having hope" and "having heart," insisting that "a certain capacity for play and joy" has to be integral to humanity's future despite the challenges of a warming planet (Haraway and Tsing 17). Haraway urges us to see "the present as a thick, complex tangle of times and places" and to prioritize our "capacities to respond" (17). Eschewing paralysis in face of the overwhelming impact of human activity on the planet, Haraway and the biologist Anna Tsing call for a new response to the historical, biological, physical, political, ethical, and religious implications of the Anthropocene, specifically inflected by non-Western, Indigenous thought.

In my capstone, discussions turned on Haraway and Tsing's sense of a "complex tangle" of lived experience. Assignments asked students to articulate new ways of thinking and being, particularly through the final project, a report on climate change that combined creative and critical research skills. Drawing on course materials, students imagined individual versions of a "human of tomorrow," a fantastical creature who would be physically and psychologically suited to a hotter, drier planet. They also extended the material covered in class by compiling an annotated bibliography of scholarly and scientific material for further reading. By adjusting our critical lens to purposely engage in fantastical thinking, we sought to exercise those responsive capacities that Haraway and Tsing find crucial for our future. Intrinsic to this assignment design is my view that speculative or fantastical thinking is not antithetical to addressing real-life issues but in fact is crucial to coming up with approaches that are truly innovative or against the grain of conventional responses (which, in the case of climate crises at least, have not worked).

The COVID-19 pandemic is only tangentially related to a warming climate, but its unprecedented impact on human life globally has forced a messy reckoning with collapsing normalcy. The pandemic highlighted the fundamental need for public health measures and social security nets in a horrifying analogue to the situation depicted in Junot Díaz's cli-fi short story "Monstro." Díaz lays out a familiar dystopic scenario in which the privileged isolate in luxury while poorer populations quickly catch a disease, "La Negrura," and are herded into camps guarded by military tanks. We discussed in class and later journaled about the last lines of the story, where the narrator speaks admiringly about his rich friend's decision to rush into the camps and document them: "[W]hat does Alex decide to do? Like an idiot he decides to commandeer one of his father's vintage burners and take a ride out to the border. . . . And what do we do, like even bigger idiots? Go with him." Using dialect-based, bleak humor, Díaz asks readers to abandon conformist ideas about linguistic propriety; with this ending, he further rejects the commonplace logic of self-preservation.[2] In the repeated phrase "like an idiot," students heard calls for qualities like bravery, intellectual curiosity, and anticolorism over the prevalent logic of self-preservation. The pandemic showed that disaster comes without warning. Our capstone dared, however sketchily, to consider the enormous and fundamental question of how to grapple with it when it arrives.

Course Specifics and the "Human of Tomorrow"

The capstone Disaster; Now What? emerged from my desire to make explicit the connections between the environment and humanistic inquiry. The course ran at my community college in the six-week winter semester, and all thirty students graduated within one or two semesters of taking the class. At my institution, most students enroll directly after high school and most transfer to a senior college after graduation (somewhat unusually for a US community college). Presuming that students would have recently encountered climate-change-related topics in high school, and given the shortness of our semester, I skipped a detailed review of the basics of climate science and spent only one week on long-form science journalism that discussed climate crises in different communities. I created a bank of recent cli-fi articles that drew on reliable academic research and connected scientific models and predictions to real-life scenarios in our own home base of New York City and around the country, from which students chose two pieces to read and annotate for homework. Hence, students were encouraged to follow their interests and curiosity to understand the extent to which climate change was already disrupting urban and rural lives across geography, class, race, culture, and other distinctions that might seem to divide one part of the country from another. In class, I introduced students, most of whom already knew youth activists like Greta Thunberg, to the work of others, like the marine biologist Ayana Elizabeth

Johnson and the journalist Elizabeth Kolbert. In the second module, we read a section from Amitav Ghosh's *The Great Derangement*, which connects climate change, industrialization, and colonialism, and a profile of the economist Thomas Piketty (Kuper), who proposes a radical mode of income taxation to pay for climate-proofing the planet. Students completing general education courses were thus invited to build upon previous courses to follow the organizing logic of our schedule.

Nonfiction readings were paired with fiction throughout. We read "Monstro" in the first module alongside the popular science articles, and the second module included Paolo Bacigalupi's "The Tamarisk Hunter" and Alexander Payne's movie *Downsizing*. The final module in this course had our most speculative works, including posthumanism as imagined by Ray Kurzweil and a long interview with Haraway and Tsing. Accompanying these were E. M. Forster's short story "The Machine Stops" and Alex Garland's movie *Annihilation*. Over the semester, we moved from realistic content addressing climate change directly toward texts that asked for freer associations and imaginative critical thinking. The materials in the first two modules focused specifically on humanity's past and how climate crises were exacerbated by human activities over the twentieth century; in the final module, *Annihilation* particularly challenged students to apply their existing intellectual equipment to imagine the "humans of tomorrow," in preparation for the final research report. Class discussions often took up the unclear, messy, and tangential connections between the later texts on our syllabus, and in a longer semester I would devote more time to this. Students in the course were already familiar with academic writing and the value of clear, logical argumentation (as in a conventional research essay); I would have liked to explore further with them the value of tangents, asides, and weak connections.

The experience of working through the movie *Annihilation* is a good example of the value that weak connections and wide parallels can have. It is not a movie about climate crises but about an alien life-form, the Shimmer, that drastically changes one section of the planet and seems to be spreading. What looks like ecological collapse to the authorities in the movie is, the work suggests, actually different modes of living on this planet. The experience of teaching this movie stays with me as a difficult but potentially useful exercise in using fantasy to practice connective thinking across different disaster scenarios. Part of my challenge lay in directing the discussion away from filmic features like characterization and plotting and toward what the movie suggests about decentering Anthropocentrism (responding to Haraway and Tsing) and imagining a posthuman future. *Annihilation* confused some students, who preferred the depiction of climate inequity in *Downsizing*, but others found it a complex thought experiment. One scene that we discussed in detail became a reference point in students' conception of a "human of tomorrow": in it, the astrophysicist Josie Radek walks into a field of spring flowers generated by the alien, Shimmer, and becomes a plantlike being. Josie acts knowing that she will change genetically into something nonhuman, an outcome the film presents as partly sad but also liberatory. Students noticed in Josie's action an echo of Díaz's narrator's "idiotic" decisions in "Monstro." The scene moved students to ponder the human aversion to change, not only in speculative fictional contexts but in practical ways also, such as in adopting green technologies or new taxation structures to ameliorate climate crises. Extrapolating from fiction, students reflected on the urgent need to question the norms that govern westernized societies, to act with more empathy for others and less greed for material possessions. They imagined "humans of tomorrow" who would be brave like Josie or Díaz's narrator. My aim in juxtaposing creative writing against scholarly research addressing real-world problems was to foster the latent connections between these works. Nonmajor courses like this capstone allow us as instructors to break down the arbitrary distinctions between different kinds of imaginative work, whether in filmmaking or economics or philosophy, and we should embrace every such opportunity with a topic as multivalent as the planet itself.

Continuing this approach, the final project asked students to design an informational report for a scholarly but nonspecialist audience, incorporating academic research and formal rhetoric into an assignment frame that emphasized the portability of our coursework beyond the classroom. Students drew on course materials to introduce the pressing importance of climate science, describe their "human of tomorrow," generate an annotated bibliography of academic research, and explain their main takeaways in a conclusion. Students used free *Canva* accounts to create professional-looking reports using available templates. Almost every report I received was exuberantly designed, using captioned images from the web and sometimes including images students had made themselves. I allocated time for a workshop on using the online design software, one unanticipated benefit of which was that students enjoyed learning to use it. A number of them said that *Canva* would be useful for making other kinds of documents or social media content. In demonstrating interplay between academic and extracurricular skill sets, the final project emphasized transferability of learning on multiple levels, asking students to demonstrate analytic, evaluative, and creative thinking in written and visual media.

Key Takeaways from Student Reflections

I will now review some of the emergent themes from a graded final self-reflection in which students examined their learning in this capstone. Over the short semester, students wrote weekly private journals; besides class participation and check-in quizzes, these journals constituted the low-stakes assignments in the course. The final report and graded self-reflection were the two high-stakes pieces that students had to complete for their final grade, and I reminded students periodically about keeping notes on course content and on their own learning process. For the final reflection, I offered guiding questions in two broad categories I had emphasized throughout the course, "content" and "academic progress":

> *Content.* It is our intention with Guttman courses that your learning will have an impact on your everyday life. How has this course prompted you to rethink ideas or ways of living that you are used to? What is one direct short-term change or amendment to your way of thinking and living that you attribute to this course and the discussions you have had with classmates, friends, and family related to it? As graduating seniors from GCC, how might your future academic career be shaped by this capstone course?
>
> *Academic progress.* It has been my intention in designing this course to help you increase the quantity of material you read, read difficult material with ease, and become adept at using and assimilating what you read to your own purposes—in other words, to encounter course content and make it yours. Think about your reading and assignments over this semester. In what ways have you succeeded in "making it yours"? Where did you encounter difficulties, and how might you address those issues in forthcoming courses?

Three broad trends emerged in their responses. First, many students said that the historical and economic readings clarified and sharpened their existing scientific knowledge about climate crises. Second, they recalled being struck by the power of individual characteristics, such as materialism or risk aversion, to profoundly affect our future success or failure as a species and by the power of philosophy to help us question anthropocentricism. Third, most students were challenged by the fictional pieces that directly and indirectly addressed human adaptations to climate crises.

I have found that students who identify as, say, science, psychology, or sociology majors sometimes resist the kinds of associative thinking that tend to be more common in humanities or literary studies courses. Most of my capstone students were from one of these three majors, which are housed in the Liberal Arts and Sciences program at our institution. In a class poll at the beginning of week 2, I asked what the relationship between speculative fiction and real-world issues might be, and only fourteen percent of the class thought "sci-fi/fantasy works are usually grounded in real life scenarios." In contrast, fifty-seven percent responded that these genres had "nothing to do with" or "only sometimes relate to" reality, and twenty-nine percent were neutral. This quick poll helped open the conversation about how imagined worlds have shaped our real lives—for instance, how sci-fi and fantasy texts have anticipated technological inventions. Forster's "The Machine Stops" was useful in this regard, because students immediately noticed that its opening, which shows people communicating with each other through screens from individual pod homes in underground cities, was speculative in 1909 when the story was published but represented a common reality in 2020. The end-of-semester reflections registered the students' surprise and, I hope, encouraged them to explore similar texts as they continue on their paths of lifelong learning.

Climate activists stress the importance of grassroots-level knowledge dissemination and discussion, but our particular semester was inconvenient for a service-learning assignment. I was pleased to learn, therefore, that some of our course content percolated beyond the classroom anyway. Anthony R. wrote with frustration, "Sometimes I find myself talking about climate change with my family but they don't understand it fully. I try to show them about it but they ignore me since they are doing their own thing." In contrast, Fenicha N. very pragmatically noted that explaining readings to her family helped her make better sense of them. Climate activism is not restricted to scientific research, and students who identified as budding photographers or filmmakers wrote about their plans to create climate change art in future. They had been bored by scientific discussions, but viewing cli-art or reading cli-fi widened their perspectives. We know that climate change is an intersectional issue that affects every aspect of lived experience, but it is not easy to connect this enormous planetary event with one's future career. Anza S. honestly summed this up:

> I always thought about becoming a doctor so I might treat people to
> the best of my abilities, as well as [have] a job that would provide me
> a sustainable income to raise a family of my own along with a higher

> social status. . . . This class made me aware that I need to step out of my
> bubble . . . and focus on the reality of our crumpling Earth through
> my future decisions and profession.

Sonia R. wrote about pursuing lab research opportunities to develop biodegradable medical gear, a comment that came back to me months later as I read about the increased volume of plastic waste generated by COVID-19 safety precautions. Stephania M. researched Ray Kurzweil's endorsement of wellness supplements because she wants to work in the wellness industry. However, she wrote that "Grossman's article [in *Time* magazine] on Kurzweil made me feel really conflicted" because it exposed the profiteering behind some industry claims. For these three science majors, the push to apply course materials to their future plans led to what they stated were unexpected sidelights into the world of work.

My students come to college with the expectation that education will directly enhance their future prospects, and our university calls itself an engine of social mobility with a mission to uplift low-income citizens into the middle classes. My classroom is a bridge space where critiques of hegemonic power and social inequality are discussed in tandem with the desire, shared across the student body, faculty, and staff, to be economically secure in the expensive city we call home. This means facing the historical realities of climate negligence by large corporations and governmental oversight bodies while also maintaining some confidence that people running those same entities will work to mitigate future planetary crises.

Notes

1. NASA currently prefers to discuss "mitigation" and "adaptation" to climate crises, since such a planetary process cannot be stopped or avoided ("Is It Too Late").

2. For a longer discussion of this story and how it can affect our teaching of literary studies, see Banerjee.

Works Cited

Annihilation. Directed by Alex Garland, Paramount Pictures, 2018.

Bacigalupi, Paolo. "The Tamarisk Hunter." *High Country News*, 26 June 2006, www.hcn.org/issues/325/tamarisk-hunter-Bacigalupi.

Banerjee, Ria. "Troubling Modernism at Community College during the Sixth Extinction." *Modernism/Modernity Print Plus*, vol. 7, no. 2, 7 Oct. 2022, https://doi.org/10.26597/mod.0233.

Díaz, Junot. "Monstro." *The New Yorker*, 28 May 2012, www.newyorker.com/magazine/2012/06/04/monstro.

Downsizing. Directed by Alexander Payne, Paramount Pictures, 2017.

Forster, E. M. "The Machine Stops." *The Eternal Moment and Other Stories*, by Forster, Mariner Books, 1970, pp. 3–38.

Ghosh, Amitav. *The Great Derangement: Climate Change and the Unthinkable.* U of Chicago P, 2016.

Grossman, Lev. "2045: The Year Man Becomes Immortal." *Time*, 10 Feb. 2011, content.time.com/time/magazine/article/0,9171,2048299,00.html.

Haraway, Donna, and Anna Tsing. "Reflections on the Plantationocene: A Conversation with Donna Haraway and Anna Tsing." Moderated by Gregg Mitman. *Edge Effects*, 18 June 2019, edgeeffects.net/wp-content/uploads/2019/06/PlantationoceneReflections_Haraway_Tsing.pdf.

"Is It Too Late to Prevent Climate Change?" *Global Climate Change: Vital Signs of the Planet*, NASA, climate.nasa.gov/faq/16/is-it-too-late-to-prevent-climate-change/. Accessed 22 June 2023.

Kuper, Simon. "This Economist Has a Radical Plan to Solve Wealth Inequality." *Wired*, 14 Apr. 2020, www.wired.co.uk/article/thomas-piketty-capital-ideology.

Kurzweil, Ray. *The Singularity Is Near: When Humans Transcend Biology.* Penguin Books, 2006.

**Rick Van Noy

Creative Responses to Climate Doom: Lessons from the Void

At a certain point in a semester of climate change literature, the mood can get dark—as perhaps it should. The news is not encouraging, and the fall-out will take a toll not only on the physical features of communities but also on the psychological topographies of individuals. While a global pandemic has seen some silver lining—carbon reduction, less noise pollution, wildlife reinhabiting the streets—it has also highlighted what climate change will ultimately do to the poor and disadvantaged, exacerbating existing social stressors and fissures.

When I traveled throughout the Southeast to talk with experts and everyday people for a book on climate change, I was inevitably asked the question, Why is an English professor interested in climate change? My answer had to do with bringing the problem down to scale, communicating it in ways that carbon dioxide measurements and hockey stick graphs were not yet doing. Too, as much as climate change is a scientific phenomenon, the reasons for inaction are political, economic, and cultural. And among the many physical adaptations we may have to make as the planet warms, as oceans expand and rise, as drought and heat intensify, and as fires and floods ravage, are the cultural ones, including coping with and

300

addressing our grief, despair, and anxiety. For that, we turn to literature, though the literature of climate change can also bring about and reinforce those concerns.

Of course, such concerns are not bad in and of themselves. Worry about the future is a sign that one cares. However, extreme anxiety can lead to shutting down, which may produce a kind of apathy. I want students to be hopeful—I don't want things to seem so dark. But I hear Greta Thunberg's voice too: "I don't want your hope. I want you to panic." Thunberg would have us act as if the world were on fire, and in some places it of course is (Workman). In addition to changes in weather and climate, fascism and nationalism are on the rise while inequalities worsen, all of it on view 24-7 through social media. Pondering the question of effective individual action can easily lead to feelings of dread and powerlessness. As my student Mary Beth Mercuri wrote in a reflection, "Imagine everything you ever worried about, magnified to a ringing in your ears that never goes away. But you can't cover your ears and ignore it, because then you are part of the problem." Robert Macfarlane has noted how the *Bureau of Linguistical Reality*, a collaborative artwork created in response to conditions of the Anthropocene, calls this "apex guilt," for how we are both responsible for and a victim of climate change. We're at the peak of human prosperity but also living at the moment when life on Earth will significantly degrade.

Conversations in my class acknowledge a common trajectory that begins with mindfulness of one's carbon footprint. People may resolve to drive less, change their diets, and make various personal sacrifices to turn back the tide of the ending of the world just a little. Yet circumstances may still require that they drive, eat food they can afford, and wear clothes made in sweatshops. They've taken steps that make their own lives harder, yet the knowledge of the planet's destruction still weighs on them, and they see, to their frustration, that people in power are laughing.

While some call these feelings climate despair, others eco-grief, the American Psychological Association uses the term *ecoanxiety* (Clayton et al.). The term may sound cute or trite, but it describes an entirely rational response to the current moment: the world is horrific and dying, and there's only so much individuals can personally do. Trapped in that futile despair, or focusing too far inward on our own actions, we lose perspective and the ability to function. This essay offers suggestions from environmental literature courses I taught in 2019 for finding alternatives to anxiety or despair when teaching climate change literature.

To Dark Mountain

When seeking a way out of the dark, it is useful to know how one came to be there. I aimed to ground students in the science of climate change by including some "no-fi" with cli-fi, including selections from Elizabeth Kolbert's *The Sixth Extinction*, chosen for its treatment of nonhuman as well as human nature. Kolbert's book makes clear the effect our putting carbon back into the air and water that has been sequestered for hundreds of millions of years is having on the planet and its species. Kolbert is a journalist, a very good one, and her purpose seems more informative than persuasive, but with the information she supplies about the five previous extinctions comes some undeniable human responsibility and "apex guilt" about the sixth.

Elizabeth Rush's *Rising* is more personal, if no less bleak, and brings together some of the information on rising sea levels and looming storms with the ways they begin to impact the author: "I began to suffer from an acute anxiety," a "gnawing uncertainty" (13). Students identified with Rush's personal struggles and their links to planetary distress, and they appreciated her attention to the plight of the Native Americans on the Isle de Jean Charles. But they found the book lacking in solutions other than organized retreat, "one of the few adaptive strategies that feels appropriately humble and, at the same time, acknowledges the scale of the threat" (249). And yet Edison Dardar, a resident of the island, told Rush (and me) he would rather not leave. He created the sign that greets visitors: "The people have a right to live where they want not where people tell them to go and live" (qtd. in Van Noy 144).

Retreat, too, is a theme in Paul Kingsnorth's *Confessions of a Recovering Environmentalist*. Kingsnorth finds contemporary environmentalism wanting an "emotional reaction to the wild world" (68). Lacking in needed ecocentrism, the current narrative views the environment as something "out there" whose problems represent an "engineering challenge"—where the summit brings not transcendence or transformation but a missed opportunity for renewable energy (79). Such "techno-utopianism" is a subset of the narrative of progress and ultimately of capitalism (241). In response to "ecocide," Kingsnorth would have us "uncivilize"; taking inspiration from Robinson Jeffers, he established a "Dark Mountain" movement in art and writing that would decenter the human in favor of something more elemental, earthy.[1]

Such stripping away characterizes Cormac McCarthy's *The Road*, a dark fictional text devoid of almost any color, save for that of fire, or life, save for an occasional mushroom (and a brook trout). When teaching this text I've often found myself thinking like its main character, "the man," picking up things along the road that might be useful, like a rock or a rusted bolt (to place on student desks), but also wary of people (which the COVID-19 pandemic intensified). Even words, such as those used in casual conversation, seem stripped away in McCarthy's text.

The Road is not expressly about climate change—the event and resulting ash appear the result of bomb fallout. But it chronicles a race to the bottom that could happen as the result of a climate breakdown. Said one mayor of a Southeast Florida community to me, "We are nice to each other now but maybe not when things go bad" (Van Noy 115). Extinction, retreat, and cannibalism can lead to a kind of Brugada syndrome, the crippling panic Elsa Bruner experiences in Nathaniel Rich's *Odds against Tomorrow*. While Elsa insulates herself from fear in a co-op garden community in Maine, her former classmate Mitchell Zukor begins to obsesses about risk, starting Future World to help businesses quantify their liability. When a tropical storm hits New York, Mitchell escapes in a canoe originally purchased as a work of art, making his way out of the devastated city and into a refugee camp. Rich seems to suggest that fear is as natural a response to our times, as human, as lust or hunger, so it deserves to be managed honestly and healthfully—not through number crunching but through creating something, including community.

Perhaps it was Rich's text, or perhaps it was Margaret Atwood's *The Year of the Flood*, that led the class to a turn toward creativity. Students no longer wanted to analyze texts; they wanted to create them. In a way, their experience is an analogue of climate change itself: the problem is known, the solutions at hand, but what are the pathways to get there? *The Year of the Flood* details the events of Atwood's previous novel, *Oryx and Crake* (in which, amid the destruction of the environment and civil society, a deadly virus developed in an elite compound wreaks havoc on humanity), from the perspective of the lower classes, specifically from the farming rooftops of the God's Gardeners ecocommune. The Gardeners are vegetarians who predict an apocalypse (which occurs) and are presented with both irony and affection. The book is filled with cruelty and violence, juxtaposed with the Blake-inspired hymns of the Gardeners, which may be read either as a parody of green zeal, hippie mysticism, and religious fervor or as something altogether serious. The Gardeners create a religion, write poetry and songs, and choose new saints, such as Rachel Carson. In her acknowledgments, Atwood invites readers to use the hymns for "amateur devotional or environmental purposes" and to visit the book's website for recordings and performance permission.

Toward Creative Community

When it came time for final projects, a weary spring class clamored for something different. They were ground down and wanted to imagine alternative scenarios, which they submitted in the form of poems, stories, a painting of the five previous extinctions, and a sea-level-rise diorama with inhabitants clinging to rooftops, clutching an American flag. One student, Mikaela Kelley, created an *Instagram* page, "VoidAid," a project she described as "winks and portents from a black hole" in the form of inspirational messages from the void. Dressed in a black spandex suit, glittery fanny pack and lilac hair, she silently impersonated a supportive consciousness that might respond to a person staring into the void, offering quotations from our class texts and elsewhere on small note cards. One read, "What will the future cost you?"—a quotation from Mitchell's Future World in *Odds against Tomorrow.* "Don't Eat Death," a reference to the vegetarian God's Gardeners, also served as a reminder of the cannibalistic *Road.* Kelley quoted Mary Oliver's essay "Upstream"—"Attention is the beginning of devotion"—to counterbalance the negative, and Walt Whitman—"a leaf of grass is no less than the journey-work of the stars"— by way of Richard Powers's *The Overstory.* On another card, she adapted McCarthy on "the absolute truth of the world"—"The cold relentless circling of the intestate earth. Darkness implacable. . . . The crushing black vacuum of the universe"—by adding the word *UNLESS,* hashtagged with McCarthy's name but also with #drseuss and #thelorax (a reference to Theodor Geisel's children's fable that harshly criticizes capitalist environmental exploitation of the environment).

Mikaela's "VoidAid" captures, in its title, a paradox at the heart of teaching climate change literature: students will stare into the dark but also seek helpful ways out. They will encounter brokenness but also want beauty. And they may need a little humor—and what can be more earthy and grounded than that?—and maybe even a little glittery light to find their way out of the darkness, off the dark mountain.

At an end-of-year celebration, my class gathered at my place around a fire ring. First, we walked, because among the best cures for anxiety is movement, indulging the flight-or-flight response, tricking the body into believing it is avoiding danger. We spied skunk cabbage ("Do you know the Oliver poem?" I prompted) and sweet flag/calamus ("Do you know the Whitman cluster?") and listened to spring peepers. We roasted hot dogs and veggie dogs, in a nod to McCarthy and Atwood. Mikaela, as the Void, brought a portable speaker to play Mary Oliver's "Gravel." And when the speaker arrived at these words, the Void repeated them to emphasize: *"dirt, mud, stars, water—I know you as if you were myself. / How could I be afraid?"*

Part of students' hunger to do creative projects seemed to come from a desire to reassert some degree of agency and control. Psychologists have long known that taking action, even symbolic action, is one of the few reliable antidotes to feelings of depression and helplessness and among the most effective ways to calm one's fears. The facts and experience of climate change can bring on feelings of powerlessness, but the climate crisis can also, paradoxically, re-empower this generation and may bind them together. Creativity will be needed, as imagination sits at the root of our challenges.

The historian Jill Lepore has written that "[d]ystopia used to be a fiction of resistance; it's become a fiction of submission . . . a fiction of helplessness and hopelessness. . . . Its only admonition is: 'Despair more.'" We cannot imagine a better future if we do not ask anyone to engage in one. Storytelling may not change the basic fact of climate change, but it can make an abstract, confusing idea more deeply felt.

Our need for connection and reciprocity looms larger in times of crisis. The world spins off its axis and we turn to those around us to keep from spiraling out along with it. I found that students couldn't function in despair, that holding on to some hope was like conserving heat in the cold. Powers has said that among other things, *The Overstory* is an antidote to loneliness: "Every form of mental despair and terror . . . seems to be related in some way to this complete alienation from everything else alive. We're deeply, existentially lonely" (John). Each of the characters in *The Overstory* experiences some form of brush with death or suicidal ideation, and then they discover one another—and trees.

When it came time to reflect on the end of a tour throughout the Southeast to learn about climate change, I was struck by the fate of the "ghost trees." These trees stand on the coast, inundated by salt water and a rising sea, tombstones of a once thriving ecosystem. They are indeed ghastly and bring on a kind of horror or grief. But I think Robin Wall Kimmerer is right that "[e]ven a wounded world holds us, giving us moments of wonder and joy." Kimmerer chooses joy over despair, "[n]ot because I have my head in the sand, but because joy is what the earth gives me daily and I must return the gift" (327).

Future World

Current college students can feel stuck in the belief that their outlook is not as bright as their parents' was at their age. My students from Appalachia, for example, have already experienced some of the wrenching trauma associated with climate change and an economic transition the rest of the country may go through. I hope to empower my students as creators of the future (or Future World).

In the future I may swap out *The Road* for Jesmyn Ward's *Salvage the Bones*, which documents the suffering and displacement of an impoverished African American community in a fictional Mississippi town. The book is full of abuse and violence, and, as Rob Nixon reminds us, climate change will inflict an accretive, incremental "slow violence." That slow violence is difficult to represent on the evening news, but Ward's novel shows it extremely well.

The next time I teach the course, I will still pursue creative projects, but I may also incorporate a project related to everyday living. I will ask students to think about Thoreau's moving to Walden Pond, or Greta Thunberg's sailing across the ocean, and to design a project on a smaller scale that changes behavior and helps communicate. Such a project might be related to how a student gets to school or work or to where their food comes from (or is). It might include involvement in a community project or a daily visit to a quiet spot. I would hope the project would bring greater awareness to how some everyday action impacts or is impacted by climate change. And I would want students to reflect on the method of persuasion from our storytellers that increased this awareness. I no longer think having a baseline of the facts is important—there will continue to be revised assessments and projections—but cultivating action and awareness is.

It is true that such personal actions can be a drop in the bucket, that we need big changes rather than small ones, and the energy companies have turned their guilt back on us. But it is a teacher's responsibility to illuminate how the small changes connect to the big ones and the local changes to the global ones.

Going after the big changes will require collective action, and collective action means realizing that no one person is holding the world on their shoulders, nor is anyone entirely powerless. In the words of my student Mercuri, "Falling into either of those extremes is the source of ecoanxiety and is what leads to a lot of good people sitting around, doing nothing, and feeling terrible about it." I do not think climate anxiety is code for people of privilege wishing to hold onto their way of life or to get "back to normal," as Sarah Jaquette Ray has suggested, as much as it is a frustration with the lack of change or progress with a new normal.

If the global pandemic taught anything, it is that people are capable of collective action. Amid all the fear, contagion, and death, I was struck by the silence too. With the abrupt halt of economic activity and traffic, some low-frequency hum ceased. Among the most googled questions is, "Are the birds louder?" (Greene). No. It's just that everything is quieter and nature is breathing again. And we are paying better attention. And when we look out and listen, the void sends back a song.

Note

1. The Dark Mountain Manifesto begins by quoting Jeffers's "Rearmament" and takes its name from the final words of the poem.

Works Cited

Atwood, Margaret. *The Year of the Flood*. Anchor Books, 2010.

Clayton, Susan, et al. *Mental Health and Our Changing Climate: Impacts, Implications, and Guidance*. American Psychological Association / Climate for Health / EcoAmerica, March 2017. PDF download.

Geisel, Theodor Seuss. *The Lorax*. Random House, 1971.

Greene, David. "Do Those Birds Sound Louder to You?" *NPR*, 6 May 2020, www.npr.org/sections/coronavirus-live-updates/2020/05/06/843271787/do-those-birds-sound-louder-to-you-an-ornithologist-says-youre-just-hearing-thin.

Jeffers, Robinson. "Rearmament." *Robinson Jeffers Association*, robinsonjeffersassociation.org/his-writing/poetry/rearmament/. Accessed 14 Aug. 2023.

———. "Vulture." *The Selected Poetry of Robinson Jeffers*, edited by Tim Hunt, Stanford UP, 2001, p. 697.

John, Emma. "Interview: Richard Powers: 'We're Completely Alienated from Everything Else Alive.'" *The Guardian*, 16 June 2018, www.theguardian.com/books/2018/jun/16/richard-powers-interview-overstory.

Kelley, Mikaela. "VoidAid: Winks and Portents from a Black Hole." *Instagram*, 29 Apr.–5 May 2019, www.instagram.com/void.aid/.

Kimmerer, Robin Wall. *Braiding Sweetgrass: Indigenous Wisdom, Scientific Knowledge, and the Teachings of Plants*. Milkweed, 2013.

Kingsnorth, Paul. *Confessions of a Recovering Environmentalist.* Graywolf, 2017.

Kolbert, Elizabeth. *The Sixth Extinction.* Picador, 2015.

Lepore, Jill. "A Golden Age for Dystopian Fiction." *The New Yorker,* 29 May 2017, www.newyorker.com/magazine/2017/06/05/a-golden-age-for -dystopian-fiction.

Macfarlane, Robert. "'Solastagia' and Other Words from Our Changing World." *On the Media,* 7 July 2017, www.wnycstudios.org/podcasts/otm/ segments/whats-in-name.

McCarthy, Cormac. *The Road.* Alfred A. Knopf, 2006.

Nixon, Rob. *Slow Violence and the Environmentalism of the Poor.* Harvard UP, 2013.

Oliver, Mary. "Gravel." *Virginia Quarterly Review,* vol. 76, no. 2, spring 2000, www.vqronline.org/gravel.

Ray, Sarah Jaquette. "Climate Anxiety Is an Overwhelmingly White Phenomenon." *Scientific American,* 21 Mar. 2021, www.scientificamerican.com/ article/the-unbearable-whiteness-of-climate-anxiety/.

Rich, Nathaniel. *Odds against Tomorrow.* Picador, 2013.

Rush, Elizabeth. *Rising: Dispatches from the New American Shore.* Milkweed, 2018.

Van Noy, Rick. *Sudden Spring: Stories of Adaptation in a Climate-Changed South.* Georgia UP, 2019.

Ward, Jesmyn. *Salvage the Bones.* Bloomsbury, 2012.

Workman, James. "Our House Is on Fire: Sixteen-Year-Old Greta Thunberg Wants Action." *World Economic Forum,* 25 Jan. 2019, www.weforum.org/ agenda/2019/01/our-house-is-on-fire-16-year-old-greta-thunberg-speaks -truth-to-power/.

Jennifer Atkinson

Stories from Our Future:
Beyond the Binary of
Climate Hope and Grief

For more than twelve years, I have asked students at the University of Washington to write about what they feel when they imagine the future. Although climate anxiety has always come up in some form, in recent years that fear seems to have hijacked students' imaginations. Here is one response from last term: "When I think about our future, all I can imagine is hell: water wars, mass extinction, infertile soil." And another: "I feel hopeless, useless, futureless, and powerless. When I imagine living through the dark times ahead, I wonder how it will feel to know we were the ones who let life on Earth to go down."

As the general public grows more aware of climate disruptions and devastations, alarm is on the rise across every age group. Indeed, a raft of new studies has even identified climate anxiety as a mental health crisis (Clayton and Karazsia; Hickman; Clayton et al.). Yet this emotional toll is especially pronounced for young people (Burke et al.; Ojala and Bengtsson). A recent survey of two thousand children and adolescents found that nearly three-quarters were deeply worried about the state of the planet and that for one in five, climate fears had disrupted sleeping or eating habits (Atherton).

All this has profound consequences for teaching and learning in environmental fields, from ecocriticism and the humanities to natural resource sciences. Many of our students hold passionate ideals and come to our programs seeking skills to make the world a better place. Yet as they learn more about the vast scale of our climate crisis and its inextricable links to other problems (such as racism, colonialism, xenophobic nationalism, inequality, and assaults on democracy), many feel despondent, guilty, and nihilistic (Ray; Hufnagel). Indeed, our curriculum itself is partly responsible for their spiraling emotions. Instructors assign gloom-and-doom materials meant to scare students into action, but as they endlessly sound the alarm on this mother of all problems, they fail to address that material's emotional toll or its impact on students' ability to respond in meaningful ways (Pihkala, "Eco-anxiety and Environmental Education"; Russell and Oakley). I remember a student who replied when asked about her postgraduation plans, "I'm going to take a year off to recover from the trauma of my courses, and then decide if I have the pain tolerance to actually do this for a living."

This emotional climate is bad for learning. As we have long known from neuroscience, chronic stress and fear trigger the "fight, flight, or freeze" response, which overrides critical and creative thinking—precisely the faculties needed to address our current crisis (LeDoux; Kelsey). Without the tools to cultivate existential tenacity, students will continue to struggle to complete our program coursework, let alone to perform the work of saving the planet they envisioned when they enrolled (Verlie; Foster).

But it doesn't have to be this way. In facing down the climate emergency, we can be honest in our presentation of the facts and our assessment of the situation while still helping students develop bold imagination, courage, and a genuine sense of agency. To successfully take up this work, educators must first identify and address three key problems feeding their students' hopelessness. These include reductive thinking about hope and grief, a sense of isolation and existential loneliness, and the inability to imagine a nonapocalyptic future. After outlining these challenges, my essay concludes with a speculative storytelling activity designed to dismantle those roadblocks and clear the space to cultivate hope, purpose, and resilience in our collective work for climate justice.

Rethinking Hope, Grief, and Purpose

In the course of giving hundreds of climate talks across North America, I have encountered tremendous diversity in the audiences, perspectives, lived experiences, and proposed solutions surrounding climate issues. Yet despite these differences, most audiences have this in common: in the question-and-answer session, someone asks, "What gives you hope?"

While there are different contexts for this question, it is clear that many people simply want to be convinced that this story will have a happy ending. They may hope I will share some exciting social or technological breakthrough—a news story they happened to miss—that promises to spare us from disaster. In short, they want to know what *others* are doing so they can go back to feeling happy.

The climate writer Diego Arguedas Ortiz has argued that nothing we learn from news or social media should make us hopeful, because "those achievements are the result of exhausted youths, overworked scientists and grieving activists. For each green bill presented in a legislative body, there's an anxious, underslept staffer who needs backup. Can one hope from the sidelines?" he asks. Such forms of "passive hope," he says, often defer action and responsibility onto others or onto the future.

Moreover, constructions of hope based on "likelihood" can be demotivating, excusing us to walk away if we feel that the odds of winning are poor (Ojala). Yet we need not limit ourselves to these mainstream understandings of hope as probability or expectation. Many climate communicators and social justice movements link hope to action—a definition that helps spur a sense of agency and obligation. Within this framework, hope is defined as something we earn or do rather than something we just "have" (Hayhoe; Macy and Johnstone).

Such a construction offers significant advantage at a moment when the need for collective action is urgent. However, "taking action" can itself lead to disillusionment when efforts fail to yield visible results. Moreover, within our grief-phobic Western culture, which urges us to steer clear of "negative" feelings or just "get over it," we may rush to action to avoid dwelling with the discomfort of climate injustice and mass extinction (Weller xxi, xix). Exclusive fixation on solutions and practical action comes at the expense of reflection, feeling, and empathetic investment—all of which are essential to the inner work needed for broader cultural change (Atkinson).

In my own teaching, I have found that reframing hope as "purpose" can be a powerful way to navigate these contradictions. A good many who work for climate justice do so not because they are convinced they will "win," or even because they feel a particular action will produce results; they do so because fighting for a livable future is the only sane and moral way to live in these times. This form of "intrinsic hope" arises from a sense of purpose and justice—values that do not depend on external factors or outcomes beyond an individual's control (Kretz; Pihkala, "Eco-anxiety, Tragedy, and Hope"). The Czech political dissident and prisoner Václav Havel famously wrote that hope is defined by the commitment to "work for something because it is good, not just because it stands a chance to succeed." For that reason, he explains, "Hope . . . is not the conviction that something will turn out well, but the certainty that something makes sense, regardless of how it turns out" (82).

In addition to reframing hope as purpose or meaning, we can deepen emotional resilience by reimagining the binary framework often used to talk about grief and hope. Climate issues invariably bring us into close proximity with suffering and loss. Yet, while it is natural to seek relief from this pain, we must not lose sight of the fact that grief is inseparable from compassion for other lives. As Malkia Devich-Cyril has written, "Joy is not the opposite of grief. Grief is the opposite of indifference. Grief is an evolutionary indicator of love—the kind of great love that guides revolutionaries."

Thus, when we try to extinguish grief, we are suppressing our love for the world. If instead we honor grief as a guide in times of adversity and loss, we open our hearts to a lasting recognition of that which we cherish most. In the process, grief amplifies and renews our commitment to fight for what we love. For this reason, climate grief carries a robust seed of hope.

Existential Loneliness

Social isolation is a second cause of climate despair. Students often say they don't feel comfortable sharing their distress with peers or friends (Hickman). Some view their social groups as indifferent; others come from families who are openly hostile to climate activism. This conclusion that no one cares feeds a painful sense of loneliness and can even cause us to question whether we ourselves are "overreacting."

Unfortunately, these feelings of social isolation tend to breed more of the same, since individuals experiencing depression or helplessness are less likely to engage with community—the very remedy needed for that alienation (Ray; brown). And beyond alleviating personal helplessness and distress, building solidarity and taking collective action is a necessary political condition for advancing climate solutions—particularly in places like the United States, where mainstream environmental messaging has long emphasized individual, sustainable lifestyle choices over collective political action.

Given the magnitude of climate change, it is essential to imagine ourselves as part of a mass movement responding in a collective fashion. Trying to tackle an emergency that is planetary in scale, that reaches far into the future, and that impacts all forms of life can make even the most passionate among us feel demoralized or conclude that one person's actions are too insignificant to matter (Moser). Yet if we see ourselves as part of a team, we remember that all contributions sync up with a larger network of change. This is what adrienne maree brown calls "emergent strategy," where a multiplicity of local interventions and connections spirals out "to create complex systems" (3). Such insights release us from the burden of trying to imagine how any one of us, as an individual, can possibly take on our climate mess.

Inability to Imagine the Future

Finally, apocalyptic narratives pose a third impediment to climate agency and hope. The first step toward creating a just and sustainable future requires, at the very minimum, the ability to imagine what such a world might look like. And yet solutions-based stories are vastly underreported while both social media and traditional news outlets overwhelmingly communicate environmental issues in a negative frame (Kelsey). For a great many of us, the dominant cultural, political, and economic institutions feel so entrenched that scholars remark, "[I]t is easier to imagine the end of the world than the end of capitalism" (Jameson 76).

To interrupt these dynamics, many climate educators invite students to examine narrative's role in imagining different futures, often drawing on Joanna Macy's framework of the three stories of our time. These include the Business as Usual story, which defends the status quo and maintains that economic growth, technological innovation, and free-market solutions will lead to prosperity for all; the story of the Great Unraveling, which predicts irreversible social and ecological collapse, mass extinction, and rising conflict over vanishing resources; and the story of the Great Turning, which sees in our current moment the possibility of profound social transformation toward genuine sustainability and environmental justice. As Macy argues, the point is not to determine which of these three stories is "correct," since they are all happening. Rather, the question is, Which one do we invest in? The third story of transition, healing, and recovery, Macy writes, is embraced by those "who know the first story is leading us to catastrophe and who refuse to let the second story have the last word" (Macy and Johnstone 5).

When we invest in a good story (like that of the Great Turning), that story can become embodied and enacted through us. Indeed, focusing on the areas where positive change is already happening and accelerating shifts our perspective, allowing us to see the current moment not as the end of the world but as the beginning of something better.

Storytelling for the Future

Humanities scholars often recognize storytelling as one of the most powerful tools for shaping how we human beings perceive our world, what is possible, and who we are or who we might become. Stories can give us visions of alternative futures and the maps to those destinations, strengthen our sense of collective purpose, and promote solidarity as we perceive ourselves acting within that story to bring about a shared goal. In this way, the right stories can help dismantle the three roadblocks outlined above while promoting intrinsic hope.

In my classes I invite students to imagine a climate-changed future where they and their communities are thriving. Students describe the unique skills and efforts they might bring to this transformation and then envision themselves as purpose-driven actors working within a wider cast of characters—including individuals, communities, and other species—all of whom are actively responding to our climate challenge. The template for this speculative story can be simple, such as a shared online document or a large-format whiteboard where participants post and view contributions.

Participants need not have a coherent vision to begin this exercise. Indeed, the collective approach invites everyone to contribute fragments of thoughts, images, and practices that feel meaningful to them and then combine and creatively reassemble those with pieces contributed by others. These fragments can be discrete scenes from a transformed future, sketches and images, lines from poems or songs, desired values, historical inspirations, and more. What motivates and ties these pieces together is the simple act of imagining a future that makes students feel empowered, inspired, and connected within a space of collective well-being.

There are countless ways such stories can be generated, and facilitators may wish to look at other models for variations on this theme. Some examples include "The Storytellers Convention," from the Work

That Reconnects Network, and *The Thing from the Future*, a game developed by the Situation Lab (see situationlab.org/project/the-thing -from-the-future/). What my version offers is a specific focus on addressing the three conditions fueling hopelessness as identified above. It is therefore important to explicitly discuss those emotional roadblocks with students as they move through this activity.

Once the group begins, the basic story emerges from four elements: Roles, Themes and Values, Transformations or Narrative Arcs, and Bridges, addressed in order as steps 1–4. As groups consult this general template, they can be encouraged to develop new components and modify the categories and "rules" to reflect specific visions and goals that resonate with them. The shared story should remain dynamic, open-ended, and participant-driven, allowing for improvisation in response to what unfolds along the way.

Roles

Participants begin by deciding which character they are within this story of a just and livable future. What specific role do they want to play, and, above all, what purpose do they wish to serve in this collective effort? They might identify as a healer, protector, teacher, gardener, storyteller, motivator, keeper of memories, or other character. Students write short bios describing their character and how they might facilitate the process of healing and repair. In addition to shifting students from the role of bystander to that of protagonist within the story, this opening move directs participants' focus away from simply hoping for a particular outcome and toward inhabiting a sense of purpose as they work to enact change. Moreover, as the unique contribution of each character takes shape alongside those of all the others, participants are reminded that they should not try to "solve" climate change in its entirety. Rather, as part of a greater cast, they can focus on a manageable intervention that draws on their personal strengths, skills, and passions (agriculture, education, healthcare, the arts, etc.).

Themes and Values

In step 2, students identify themes that would serve as guiding principles within this story of collective transformation. What values should be centered (e.g., justice, biodiversity, slowness, wonder, community)? Once all contributions have been submitted, they can be combined and shared as a word cloud or represented in other creative ways. Having identified values that are most meaningful to the group, participants can also return to step 1 to see if their role or purpose allows them to fully inhabit these core values. Storytellers also look ahead to step 3 to ensure that their envisioned transformations are advancing the underlying principles articulated in this step. Completing this step before developing narrative arcs helps avoid the trap of a predetermined outcome in which participants simply reproduce established patterns and institutions from our present circumstances. Rather than beginning with a final outcome (e.g., "Here's what agriculture looks like" or "Here's a master plan for our childcare system"), they begin with desired values and then think through the actions and conditions needed to promote those principles.

At this point in the exercise, I remind students that their most meaningful insights may emerge from personal and collective experiences of struggle and loss. They reflect on a particular source of grief or anger (such as a displaced community, an extinct species, or a destabilized weather system). How does this loss make them feel, and what is their pain showing them? What commitments might we reinforce and bring into the center of the collaborative story as a result (gratitude, openness to honoring our pain, etc.)? How is their particular memory or insight kept alive and honored in the wake of the loss? In addition to acknowledging the pain inherent in our climate situation, these reflections help students recognize the wisdom and compassion that can arise from grief and loss.

Transformations or Narrative Arcs

Step 3 asks students to describe concrete changes that must occur to create their desired future. What must be dismantled, reimagined, and rebuilt? Here, each student describes a scene, an episode, a moment, or an event, and those pieces are combined into a larger story. Some students have combined their narratives into a story of a gathering or festival held at a community garden; others have set their story at a future college that is free and accessible to all, where students gain hands-on experience learning about climate justice in a wetland restoration project with the local community. Enlivening these scenes with explicit, multisensory detail (describing shapes, colors, sounds, smells, temperature), character names, and spoken exchanges helps animate the "feel" of this possible future. This vivid specificity stimulates creative faculties as well as the conviction that what we describe is really possible. As Macy and Johnstone write, research indicates that people who engage a problem by imagining it has already been solved "are more creative and detailed in inventing potential solutions" (172).

While each participant's scene may only be a small piece of the puzzle, those single contributions now feel less "inadequate" as they are viewed as part of the larger constellation. To that end, participants are invited to identify connections and synergies between different story components. How might these episodes add up to a larger story of transformation? What are some cause-and-effect relations between them? What conflicts might arise between these various contributions, and how might those tensions be creatively addressed?

Bridges

After completing steps 1–3, participants use the remainder of the course term to look for real points of emergence in the present that connect to elements of the future they just imagined. What movements, actors, efforts, and developments are currently underway and already working to create that story of transformation? How might students get involved to support and accelerate that change, and how might they inhabit the "role" they created for themselves right now? Which organizations or efforts need the specific skills they identified with their character? How might those steps activate their sense of purpose in the near future or present moment?

Perhaps the key takeaway of this storytelling activity is a sense of solidarity. Often, the source of our personal despair isn't the crisis itself so much as a longing for fellowship and a sense of purpose. Paradoxically, the meaning and connection we crave can be deepened by adversity (Solnit). In assessments following this activity, students repeatedly say that imagining themselves as part of a historic movement rising up to confront and repair planetary injustice brought them a sense of joy, purpose, and even adventure.

They also know, however, that they have not left this experience with a precise map or preview of what will actually happen. Rather, what they take away is a set of mental images and emotional tools: inspiration, courage, excitement, and hope rooted in the possibility of contributing to the work ahead. The details and odds may remain obscure, but, as Rebecca Solnit reminds us, the darkness charactering our future need not be felt as bad or gloomy but can be understood simply as the murky unknowability of an uncertain path that could go multiple ways. "Hope," she writes, "locates itself in the premises that we don't know what will happen and that in the spaciousness of uncertainty is room to act." Telling collective stories that redirect our despair is an essential first step in building the vision and courage to take that transformative action.

Works Cited

Atherton, Richard. "Climate Anxiety: Survey for BBC Newsround Shows Children Losing Sleep over Climate Change and the Environment." *BBC*, 3 Mar. 2020, www.bbc.co.uk/newsround/51451737.

Atkinson, Jennifer. "Mourning Climate Loss: Ritual and Collective Grief in the Age of Crisis." *CSPA Quarterly*, vol. 32, spring 2021, pp. 9–19.

brown, adrienne maree. *Emergent Strategy*. AK Press, 2017.

Burke, Susie, et al. "The Psychological Effects of Climate Change on Children." *Current Psychiatry Reports*, vol. 20, no. 35, 2018, http://doi.org/10.1007/s11920-018-0896-9.

Clayton, Susan, and Bryan Karazsia. "Development and Validation of a Measure of Climate Change Anxiety." *Journal of Environmental Psychology*, vol. 69, 2020, http://doi.org/10.1016/j.jenvp.2020.101434.

Clayton, Susan, et al. *Mental Health and Our Changing Climate: Impacts, Implications, and Guidance*. American Psychological Association / Climate for Health / EcoAmerica, 2017. PDF download.

Devich-Cyril, Malkia. "Grief Belongs in Social Movements. Can We Embrace It?" *In These Times*, 28 July 2021, www.inthesetimes.com/article/freedom-grief-healing-death-liberation-movements.

Foster, John. *After Sustainability: Denial, Hope, Retrieval*. Routledge, 2015.

Havel, Václav. *Disturbing the Peace: A Conversation with Karel Huizdala*. Vintage, 1991.

Hayhoe, Katharine. "In the Face of Climate Change, We Must Act So That We Can Feel Hopeful—Not the Other Way Around." *Time*, 12 Aug. 2021, time.com/6089999/climate-change-hope/.

Hickman, Caroline. "We Need to (Find a Way to) Talk about . . . Eco-anxiety." *Journal of Social Work Practice*, vol. 34, no. 4, 2020, pp. 411–24, http://doi.org/10.1080/02650533.2020.1844166.

Hufnagel, Elizabeth. "Attending to Emotional Expressions about Climate Change: A Framework for Teaching and Learning." *Teaching and Learning about Climate Change*, edited by Daniel Shepardson et al., Routledge, 2017, pp. 43–55.

Jameson, Fredric. "Future City." *New Left Review*, no. 21, May-June 2003, pp. 65–79.

Kelsey, Elin. *Hope Matters: Why Changing the Way We Think Is Critical to Solving the Environmental Crisis*. Greystone Books, 2020.

Kretz, Lisa. "Hope in Environmental Philosophy." *Journal of Agricultural and Environmental Ethics*, vol. 26, 2013, pp. 925–44.

LeDoux, Joseph. *Anxious: Using the Brain to Understand and Treat Fear and Anxiety.* Penguin Books, 2016.

Macy, Joanna, and Chris Johnstone. *Active Hope: How to Face the Mess We're in without Going Crazy.* New World Library, 2012.

Moser, Susanne C. "The Work after 'It's Too Late' (to Prevent Dangerous Climate Change)." *WIREs Climate Change,* 23 Oct. 2019, http://doi.org/10.1002/wcc.606.

Ojala, Maria. "Hope and Anticipation in Education for a Sustainable Future." *Futures,* vol. 94, 2017, pp. 76–84.

Ojala, Maria, and Hans Bengtsson. "Young People's Coping Strategies Concerning Climate Change: Relations to Perceived Communication with Parents and Friends and Proenvironmental Behavior." *Environment and Behavior,* vol. 51, no. 8, 2018, pp. 907–35, http://doi.org/10.1177/0013916518763894.

Ortiz, Diego Arguedas. "Is It Wrong to Be Hopeful about Climate Change?" *BBC,* 9 Jan. 2020, www.bbc.com/future/article/20200109-is-it-wrong-to-be-hopeful-about-climate-change.

Pihkala, Panu. "Eco-anxiety and Environmental Education." *Sustainability,* vol. 12, no. 23, 2020, http://doi.org/10.3390/su122310149.

———. "Eco-anxiety, Tragedy, and Hope: Psychological and Spiritual Dimensions of Climate Change." *Zygon,* vol. 53, no. 2, 2018, pp. 545–69.

Ray, Sarah Jaquette. *A Field Guide to Climate Anxiety.* U of California P, 2020.

Russell, Connie, and Jan Oakley. "Engaging the Emotional Dimensions of Environmental Education." *Canadian Journal of Environmental Education,* vol. 21, 2016, pp. 13–22.

Solnit, Rebecca. "Hope Is an Embrace of the Unknown: Rebecca Solnit on Living in Dark Times." *The Guardian,* 15 July 2016, www.theguardian.com/books/2016/jul/15/rebecca-solnit-hope-in-the-dark-new-essay-embrace-unknown.

"The Storytellers Convention." *Work That Reconnects Network,* 1 Dec. 2017, workthatreconnects.org/resource/the-storytellers-convention/.

Verlie, Blanche. *Learning to Live with Climate Change.* Routledge, 2021.

Weller, Francis. *The Wild Edge of Sorrow: Rituals of Renewal and the Sacred Work of Grief.* North Atlantic Books, 2015.

Sarah Jaquette Ray

Afterword:
The Urgency of Slow Teaching

Even before the COVID-19 pandemic, many of my students were burning out from the sheer labor of staying housed, fed, and upright amid the flood of bad news they were learning in school and from the media. More than ever, students were precarious economically, nutritionally, and emotionally. Add to their woes the snowballing crises of climate injustice and their growing sense that climate change was becoming the story of their lives: if they weren't horrified by what they witnessed happening in the most vulnerable places and to those least responsible for the crisis, they worried about climate change knocking on their own doors.

Then the pandemic struck, just at a time when many young people were already suffering a whole lot. I thought they were unwell before, but the pandemic wiped the life from them. I'm writing these sentences at a low point in my career as a college professor; the empty hallways and *Zoom*-room black boxes reflect such an intense void in my students, I hardly know what to do. I can't get a pulse from them. They're scattered to the winds of grief and overwhelm.

"Global dread," "solastalgia" (Albrecht et al.), ecoanxiety, and climate trauma are on the rise, most strikingly among the college-aged (Hickman et al.). And it's only going to get worse as climate change accelerates these myriad front lines of suffering and those in power continue to deflect responsibility. At the UN Climate Change Conference in Glasgow (COP26) in 2021, the United States and China—the world's worst emitters—couldn't find common ground or muster the leadership needed to turn this train wreck around. Young people are furious, they feel betrayed, and they can't understand why everyone else is acting as if their house is *not* on fire.

In ditching school to participate in strikes for #FridaysforFuture, Greta Thunberg and millions of young people around the world raised the question, What good is going to school and preparing for a career while the planet is dying? Educators everywhere are hearing them. What does education need to be in this moment? Given students' sense of urgency about the viability of life on this planet—including their own—why should they come to our classes? Is it ethical to keep teaching the way we always have? How can we attune our pedagogy to honor their anguish and help them face the challenges that scientists forecast will unfold in their lifetimes?

In this context, it is surreal to read the essays in this collection. With each essay, my wonder intensified: Who are these students who can sit still long enough to read a novel (much less multiple novels), to have a discussion in class about what climate fiction is, to do close readings, and to imagine their futures? Where are the stories of students breaking down in classrooms, hallways, offices, and assignments? Why aren't they enraged or despondent to the point of storming out of their sterile classrooms? Why aren't they protesting in the name of *urgent action, now*?

Magdalena Mączyńska summarizes this dilemma poignantly in her contribution, and I identify with her inner turmoil:

> Is it my role to introduce students to the basic science of climate change? . . . Or is my first duty to emphasize the social justice implications of the climate crisis? Is it to foster activism? to address climate anxieties? or to field urgent questions: "Is it too late?" and "What can I do to help?" Where among these pressing matters do I find room for the slow, meticulous work of literary analysis?

What I came to realize as I myself slowed down my experience of reading the contributions, as I let the ideas speak, and as I imagined these educators agonizing over these very questions was that they care so deeply and are channeling that care to defining a pedagogy that can meet this moment—a pedagogy that insists on the urgency of slow teaching.

In this volume, I met students who read novels and poetry (*not real enough!*). I met students learning the craft of close reading and the politics of canonicity (*not useful enough!*). I meandered with them through the gentle slopes of genres and tropes (*not relevant enough!*). With them, I devoted myself like a monk to the transcendent practice of textual analysis (*not solution-based enough!*). These students accept fiction as a meaningful approach to the climate crisis and are conditioning their spirits in service to a world they desire.

In truth, they are doing exactly what the climate emergency needs us all to do: not setting up survivalist camps under freeways or becoming hermits in caves but undertaking slow, thoughtful deliberation—in conversation with literature—about how to do less harm and how to live in the face of disaster and injustice. The educators herein are beacons for us, illuminating ways to navigate teaching in tumultuous times and retool our humanities training for the existential gravity of what our students (and often our own children) are facing. These contributors are themselves grieving and anxious, yet they refuse hopelessness, and they are doing everything they can to ensure their students do too. What comes across in these testimonies of tender teaching is that the authors deeply, profoundly care—about climate justice, the more-than-human world, and their students. There are no easy fixes to what we face, and this volume is testament to the time and energy it takes to figure out what we can and must do anyway.

Care itself is never enough. It can be expressed in all kinds of ways—some that are in service to the protection of the object of care and some that are not. Humans are flawed this way; sometimes a care act is a reactive impulse caused by a hijacked amygdala—the "fight, flight, or freeze response"—preventing the thinking mind from helping us select the most effective course of action. When we receive a scary stimulus, such as the constant news about climate catastrophes and human suffering everywhere, most of us react impulsively and do not have the training required to force a pause to think through what is real, what the best course of action is, who needs to be involved, and how the action may or may not reduce harm.

This is why slowing down in this moment of such extreme urgency is precisely the training we need. As Bayo Akomolafe and Marta Benavides write, "The times are urgent. We must slow down." Care enlisted in service to that which we care about requires time. And while we may think we don't have enough time, this is an illusion. We have exactly the time of our lifespans. That's all anybody ever has. Perpetually reacting from heightened anxiety, fear, or anger can leave us in a state of chronic stress, trauma, and burnout. It fragments our efforts, depletes our capacity for effective action, and frays collective bonds.

How do the slow, careful practices of reading and discussing novels save the planet? One example is Parker Krieg's historical contextualization of college students as the "professional-managerial class," which helps explain their cognitive dissonance as at least partially a result of the tension they feel between serving capital and wanting to serve the planet. With this awareness, the sense of inner conflict can be lessened and students can creatively imagine how to put their values into action.

Another example is the way the volume centers justice in its pedagogy. Several pieces scrutinize urgency as a mode: it can be an excuse to override voices and epistemologies that aren't the loudest or are silenced by the din of dominant ideology. The fetishization of climate fiction in recent years raises the question of whether the genre itself reifies a white environmental imaginary, one that leaves green colonialism intact. If we are to do the work of climate justice, not just glorify the ability of some people to buy their way out of the worst effects of climate change, then a deliberative process of asking whose voices register in the conversation and how genres, texts, and other formal codes open or foreclose that visibility is needed. This takes time. Critical environmental justice pedagogy, as April Anson describes it, is a painstaking endeavor: anti-racist pedagogical practices demand that educators "encourage reflexivity, prepare for and welcome difficulty, meet students where they are, engage affective and embodied dimensions of learning, and build learning communities." What happens if we choose not to take the time to do this, in the name of utility? I dread to think.

In defining climate injustice as a crisis of attention—of how those in relative power notice, see, and thereby care about the impacts of their actions (both positive and negative) on a wide range of vibrant life—several authors describe how they train students to notice and attune themselves to the world around them. If we don't know how to see the world, how can we see climate change, much less generate effective action? Unlearning a capitalist training and relearning the art of attention truly takes time.

The unglamorous drudgery of building a world different from the dystopia we fear is a profound act of resistance to that seemingly inevitable fate. Many students accept the forecasts as truth and are never taught any other story. In a challenge to that negativity bias, speculative fiction pulls students in the direction of imagining how things could be instead—the first step required in building a world that avoids the worst of those forecasts. Can students both recognize urgency and also cultivate the creativity and efficacy required to imagine a future they desire? As Tobias Menely concludes, speculative fiction "offers students a chance to collectively imagine the future in a way that acknowledges necessity while making room for contingency." Imagining a better world requires a spaciousness of spirit. It can't be rushed by fears of apocalypse on students' doorsteps.

Another theme throughout these pages is a pedagogy of collectivity in place of the individualism that is otherwise baked into higher education. If our crisis is one of disconnection and alienation from others, including the more-than-human world, how can kinship ties be repaired? What is the work of relation? How long does it take?

This slow work of critique also supports another kind of interconnection—the kind that systems thinking helps us analyze toward more effective solutions. Systems thinking approaches, as evidenced in many essays in this collection, "connect struggles that have too often been presented as discrete from one another." Ben Jamieson Stanley and Emily S. Davis point out that environmental humanities classrooms are the perfect place for "historicizing and connecting struggles for racial justice and climate change," because they "enable students to perceive and tell new stories about what has happened, what might happen, and what we can seek to actualize instead."

Because climate change resists risk perception, it has been a humdinger to narrativize, as many scholars have argued (Nixon; Ghosh), but the slow work of literary analysis helps us connect dots between what we can see and what we cannot see, as Patrick Whitmarsh beautifully explains: "not all of us see climate change in the world around us; or, rather, we do not realize that we see it. We may see weather, construction, cars, concrete, infrastructure, trees, flowers, and birds whose presence and behavior are part of climate change, but they may not speak to us as climate change." According to Whitmarsh and others in this volume, students need to learn the process of "derealization" rather than the "realism" of the protest chants of the mainstream climate movement. Going against this mainstream view of climate realism is hard, and so it takes time.

Connecting dots is reason enough to slow down and think about systems before rushing in with overly simple solutions to problems right in front of us. Putting out fires as they crop up becomes an exercise in reactivity; without deeper analysis, we can provide only Band-Aid solutions to problems that require deeper healing. This is why bodies such as FEMA are not in themselves a solution to natural disasters; we need time and intellectual capacity to bring systems thinking to bear on crafting preventive policy, community-building, and just processes.

These are just a few of the ways that the essays demonstrate the need for a kind of patient critique: experimenting with various lenses (or "angling," as Barbara Leckie calls it), tending to that which is hard to see, developing new senses, challenging existing epistemologies and exploring better ones, and fortifying relationships wherever possible.

It gives me great relief and hope that so many wise, brilliant educators are slowing students down long enough to do this work. In doing so, they repeatedly amplify the essential role of the environmental humanities in engaging climate change. They amplify the voices of their students—the generation who, even at the level of greatest privilege, will be saddled with the worst of climate change's unfolding. In these essays, students and their guides face their emotions about all this and use them to shine light on what they love.

Indeed, the effect is a pedagogy of sorrow/joy, a theme that comes up in several essays. Anson quotes Ross Gay's poetic exploration of joy, which he sees as sorrows joining:

> Is sorrow the true wild?
> And if it is—and if we join them—your wild to mine—what's that?
>
> . . . .
>
> I'm saying: What if that is joy? (49–50)

And Rick Van Noy cites Robin Wall Kimmerer, who chooses joy over despair, "[n]ot because I have my head in the sand, but because joy is what the earth gives me daily and I must return the gift."

This paradoxical work of holding multiple truths—joy/sorrow, grief/love, urgency/care, action/rest, reality/unreality, future/past, macrocosm/microcosm, power/humility, to name just a few—seems to me the most crucial work we can do as educators in navigating the myriad tendrils of how we and our students will (and already do) experience the climate crisis.

We do need to act quickly, but what will happen if we fail to take care in doing so? How can we help students reconcile the tension they feel between urgency and intimate attention? The failure to slow down long enough to notice the scariness of all this and identify the best courses of action, much less to engage in the intellectual critique required to imagine fresh courses of action, is arguably what caused the crisis in the first place.

Works Cited

Akomolafe, Bayo, and Marta Benavides. "The Times Are Urgent: Let's Slow Down." *Báyò Akómoláfé*, www.bayoakomolafe.net/post/the-times-are -urgent-lets-slow-down. Accessed 7 July 2023.

Albrecht, Glenn, et al. "Solastalgia: The Distress Caused by Environmental Change." *Australasian Psychiatry*, vol. 15, no. 1, supp. 1, Feb. 2007, pp. S95–S98, https://doi.org/10.1080/10398560701701288.

Gay, Ross. *The Book of Delights.* Algonquin Books, 2019.

Ghosh, Amitav. *The Great Derangement: Climate Change and the Unthinkable.* U of Chicago P, 2016.

Hickman, Caroline, et al. "Climate Anxiety in Children and Young People and Their Beliefs about Government Responses to Climate Change: A Global Survey." *Lancet Planet Health*, vol. 5, no. 12, Dec. 2021, e863–e873, https://doi.org/10.1016/S2542-5196(21)00278-3.

Nixon, Rob. *Slow Violence and the Environmentalism of the Poor.* Harvard UP, 2013.

Notes on Contributors

Sofia Ahlberg is vice dean of education and public engagement in the Faculty of Languages, Uppsala University. She is the author of *Teaching Literature in Times of Crisis* (2021) and is currently investigating transformational reading practices and magic to help people understand the agency of the literary imagination when faced with apparently insurmountable obstacles.

Melissa Anderson is associate professor in Hannon Library at Southern Oregon University. Her research interests include nineteenth- and twentieth-century literatures in English, the history of science, and the teaching of writing and information literacy. She has published essays on nineteenth-century literature and on information literacy instruction and is currently working on a book about integrating information literacy instruction in English literature courses.

April Anson is assistant professor of English at the University of Connecticut, where she works at the intersection of environmental humanities, Indigenous American studies, and political theory. Previously, she was assistant professor of public humanities at San Diego State University and a Mellon postdoctoral fellow at the University of Pennsylvania. Her research uses literary analysis to trace the historical and ongoing relationship between climate change, white supremacy, and American environmental thought as well as the Indigenous American environmental justice traditions that eclipse those relations. Anson is a cofounder of the Anti-Creep Climate Collective and coauthor of *Against the Ecofascist Creep*, and her research has appeared in *Boundary 2*, *Resilience*, *Environmental History*, *Western American Literature*, and elsewhere.

Jennifer Atkinson is associate professor of environmental humanities at the University of Washington, Bothell. Her seminars on ecogrief and climate anxiety have been featured in *The New York Times*, *National Geographic*, NBC News, *The Washington Post*, *Grist*, and many other outlets. Her podcast *Facing It* also provides tools to channel ecoanxiety into action. Atkinson is the author of *Gardenland: Nature, Fantasy and Everyday Practice* (2018) and editor of *An Existential Toolkit for Climate Justice Educators*, which offers strategies to help students and educators navigate the emotional toll of climate breakdown.

Nassim W. Balestrini is professor of American studies and intermediality at the University of Graz and head of the Centre for Intermediality Studies in Graz. Previously, she taught at the universities of Mainz, Paderborn, and Regensburg (in Germany) and at the University of California, Davis. Her research addresses intermediality and adaptation theories, life writing across media, hip-hop (particularly rap as poetry and contemporary Indigenous and Alaskan artists), African American literature, borders and mobility, and contemporary theater and poetry, including climate change theater. In addition to having published multiple articles on Chantal Bilodeau and on climate change microdramas, she has been exploring ways of collaborating with artists in order to integrate creative work into literary and cultural studies classes.

Ria Banerjee is associate professor of English at Guttman Community College and a consortial faculty member in film studies at the Graduate Center, City University of New York. Her scholarly interests are in British and European modernism and post–World War II film. She has written for *Modernism/Modernity* and the *South Atlantic Review*, among other scholarly venues, and is currently at work on a monograph on spatiality in interwar British fiction. Banerjee teaches undergraduate courses in writing, literature, and media studies and graduate courses in film studies.

Ali Brox is assistant teaching professor in the Environmental Studies Program at the University of Kansas. She specializes in the environmental humanities, and her research and teaching interests include the intersection between ecocriticism and postcolonial theory with particular focus on environmental justice discourse. Her work has appeared in *Green Humanities Journal*, the *Journal of the Midwest Modern Language Association*, *Utopian Studies*, and *Studies in American Humor*, and she is a sub-awardee on a multi-institutional National Science Foundation grant focused on evaluating student knowledge of complex systems in environmental education.

Brianna R. Burke is associate professor of environmental humanities and American Indian studies in the English department at Iowa State University. Currently, she is working on a book titled "Becoming Beast: The Humanimal in Climate Justice Literatures," which explores the morphing ideology of what it means to be (considered) animal in a world of decreasing resources, and her work explores the connections between environmental justice, American Indian studies, climate change, and multispecies kinship. Her most recent publication on conjoined corporeal precarity, titled "Disemboweling the Hyperreal in Bong Joon-ho's *Okja*," is forthcoming in *Animal Futurity: A Speculative Exploration of the Future of Human-Animal Relations*.

Matt Burkhart serves as a lecturer in the SAGES program at Case Western Reserve University, where he teaches writing-intensive seminars on urban environmentalism and climate change fiction. In addition to this and earlier essays in the fields of regional and environmental humanities, he has recently published "Trees Are Better Than Stone: Vital Commemoration in Octavia Butler's *Parable* Novels" in a special issue of *Western American Literature* focusing on California-based cli-fi. That essay is part of a monograph in progress, provisionally titled "Anthropocenotaphs: Structures of Mourning and Resilience in the Anthropocene."

Stef Craps is professor of English literature at Ghent University, where he directs the Cultural Memory Studies Initiative. His research interests lie in twentieth-century and contemporary literature and culture, memory and trauma studies, postcolonial theory, and ecocriticism and the environmental humanities. He is the author of *Postcolonial Witnessing: Trauma Out of Bounds* (2013) and *Trauma and Ethics in the Novels of Graham Swift: No Short-cuts to Salvation* (2005), a coauthor of the New Critical Idiom volume *Trauma* (2020), and a coeditor of *Memory Unbound: Tracing the Dynamics of Memory Studies* (2017). He has also edited or coedited special issues of journals, including *American Imago*, *Studies in the Novel*, and *Criticism*, on topics such as ecological grief, climate change fiction, and transcultural Holocaust memory.

Emily S. Davis is associate professor of English and women and gender studies at the University of Delaware. She is the author of *Rethinking the Romance Genre: Global Intimacies in Contemporary Literary and Visual Cultures* (2013) as well as articles on South African fiction, queer neoliberalism, and human rights.

Clare Echterling is a lecturer in English at Caldwell University. Her research and teaching on environmental literature focus on imperialism and environmental justice, especially in children's and young adult literature. Her work appears or is forthcoming in *The Oxford Handbook of Young Adult Literature, Children's Literature, Children's Literature in Education*, and elsewhere.

Christina Gerhardt is Barron Professor of Environment and the Humanities at the High Meadows Environmental Institute at Princeton University, associate professor at the University of Hawai'i at Manoa, and a senior fellow at the University of California, Berkeley. She has been awarded fellowships by the Fulbright Commission, the National Endowment for the Humanities, the Newberry Library, and the Rachel Carson Center and held visiting appointments at Columbia University, the Free University in Berlin, and Harvard University. She is editor of *ISLE: Interdisciplinary Studies of Literature and Environment*. Her most recent book is *Sea Change: An Atlas of Islands* (2023).

Teresa A. Goddu is professor of English at Vanderbilt University. She is the author of *Gothic America: Narrative, History, and Nation* (1997) and *Selling Antislavery: Abolition and Mass Media in Antebellum America* (2020). Her work has appeared in *American Literary History, Book History, MELUS, African American Review, Common-Place, South Atlantic Quarterly, Studies in American Fiction,* and other venues. She is the recipient of two grants from the National Endowment for the Humanities and a Senior Specialist Fulbright award. Her current research focuses on the environmental humanities, specifically contemporary US climate fiction.

Mary Ann Gosser-Esquilín is director of the university honors program and professor of Spanish and comparative literature in the Department of Languages, Linguistics, and Comparative Literature at Florida Atlantic University. Her areas of interest and research encompass the literatures of the French- and Spanish-speaking Caribbean. She has taught at Rutgers University; at Denis Diderot, Université de Paris VII; and at the University of the West Indies, Mona Campus, Jamaica, as a visiting Fulbright scholar. Her publications include *Culture, Nature, and the Other in Caribbean Literature: An Ecocritical Approach* (2023) and articles in *Sargasso, CLA Journal, Confluencia, New Mango Season, Centro Journal,* and *CUALLI: Latin American and Iberian Food Studies Review.*

Andrew Hageman is associate professor of English and director of the Center for Ethics and Public Engagement at Luther College in Decorah, Iowa. He researches and teaches intersections of technology, ecology, and ideology and has published on an eclectic set of topics, including David Lynch's cinema, Chinese science fiction, blockchain and the social imaginary, and Björk's *Biophilia*. He contributed an essay to *Ecocinema Theory and Practice* (2013); coedited *Global Weirding*, a special issue of *Paradoxa* (2016); and cowrote, with Regina Kanyu Wang, a chapter on science fiction ecocinema for *Ecocinema Theory and Practice 2.*

Thomas Hallock is professor of English and Florida studies at the University of South Florida, where he teaches on the St. Petersburg campus. His publications include *From the Fallen Tree: Frontier Narratives, Environmental Politics, and the Roots of a National Pastoral* (2003), *William Bartram, the Search for Nature's Design: Selected Art, Letters, and Unpublished Writings* (2010), and, most recently, *A Road Course in Early American Literature: Travel and Teaching from Atzlán to Amherst* (2021).

Jeffrey Johansen, professor emeritus of biology at John Carroll University, is an internationally recognized researcher in cyanobacterial taxonomy and systematics, with collaborators in the Czech Republic, Mexico, Russia, and

Brazil. He is an associate editor for two academic journals and serves on the editorial board for the *Journal of Phycology*. Johansen has been recognized by John Carroll University as a recipient of the Distinguished Faculty Award and received an honorary doctoral degree from the University of South Bohemia. His areas of expertise include cyanobacterial taxonomy and molecular systematics, diatom taxonomy and ecology, water quality, and biological soil crust floristics and ecology.

Parker Krieg teaches exploratory and interdisciplinary studies at the University of Nebraska, Omaha. He was previously a lecturer in the global studies program at the University of Nebraska, Lincoln, and held a postdoctoral fellowship in environmental humanities at the University of Helsinki, affiliated with the Helsinki Institute of Sustainability Science. He is coeditor of *Situating Sustainability* (2021), and his recent article, "Beyond Flexible Nature: Raymond Williams after Post-Fordism," appears in the Raymond Williams centenary issue of *Coils of the Serpent: Journal for the Study of Contemporary Power*.

Hannah Kroonblawd is assistant professor and director of creative writing at Malone University. Her poetic and critical interests include poetics in the Anthropocene, ecotheology and eschatology, and creative writing pedagogy. She is author of the chapbook *The Leafcutters, the Minor Saints* (2021), and her writing can be found in a variety of literary magazines and journals.

Barbara Leckie is professor in the Department of English and the Institute for the Comparative Study of Literature, Art, and Culture at Carleton University. She is the author of *Open Houses: Poverty, the Architectural Idea, and the Novel in Nineteenth-Century Britain* (2018) and *Climate Change, Interrupted* (2022), among other books and articles. She is also the editor of *Sanitary Reform in Victorian Britain: End of Century Assessments and New Directions* (2013); coeditor, with Janice Schroeder, of a new edition of Henry Mayhew's *London Labour and the London Poor* (2019); the founder of the Carleton Climate Commons; and the academic director of Re.Climate: Centre for Climate Communication and Engagement.

Magdalena Mączyńska is professor of English and world literatures at Marymount Manhattan College in New York City, where she teaches courses in climate fiction, contemporary anglophone fiction, literary theory, and academic writing. Her research interests include climate fiction; intersectional ecocritical theory; postmodernist, postcolonial, and postsecular fiction; urban theory and creative practice; and critical pedagogy. Her work on contemporary urban and climate fiction has been published in *Contemporary Literature, ISLE: Interdisciplinary Studies in Literature and Environment,*

Frame, and the *Journal of Modern Literature*. She is the author of *The Gospel According to the Novelist: Religious Scripture and Contemporary Fiction* (2015).

Ted Martinez is a teaching professor at Northern Arizona University (NAU), where he teaches classes on plants and the literature of climate change in the NAU Honors College. Every other year he directs NAU's Grand Canyon Semester, an immersive, semester-long experiential learning program centered on interdisciplinary study of the Grand Canyon and Colorado Plateau ecoregion. He is a founding editor of *Carbon Copy*, an online journal at the intersection of climate science and literary arts that seeks to promote interdisciplinary discussion on climate change.

Robert P. Marzec is professor of environmental and postcolonial studies in the Department of English at Purdue University. He is the author of *Militarizing the Environment: Climate Change and the Security Society* (2015) and *An Ecological and Postcolonial Study of Literature* (2007). He is editor of *MFS: Modern Fiction Studies* and has published articles in *Boundary 2, Radical History Review, Public Culture, Postmodern Culture*, and *Elementa: Science of the Anthropocene*.

Tobias Menely, professor of English at the University of California, Davis, is the author of *The Animal Claim: Sensibility and the Creaturely Voice* (2015) and *Climate and the Making of Worlds: Towards a Geohistorical Poetics* (2021).

Jason de Lara Molesky is a Mahindra postdoctoral fellow in the environmental humanities at Harvard University and, starting in 2024, assistant professor of English at Saint Louis University. His book project examines the lifeworlds of global American company towns through studies of contemporary art and literature. His essays have appeared or are forthcoming in *Modern Fiction Studies, American Literature, The Georgia Review*, and elsewhere. He is coeditor of *Cli-Fi and Class* (2023) and of *Teaching Energy Humanities*, in progress.

Jo Alyson Parker is professor emerita of English at Saint Joseph's University, where she has taught courses in narrative and time, literary theory, and the eighteenth- and early-nineteenth-century novel. Her publications include *The Author's Inheritance: Henry Fielding, Jane Austen, and the Establishment of the Novel* (1998) and *Narrative Form and Chaos Theory in Sterne, Proust, Woolf, and Faulkner* (2007). She has also published essays dealing with narrative, time, and memory in the work of such writers as Kate Atkinson, Ted Chiang, J. M. Coetzee, Shelley Jackson, David Mitchell, and

Tom Stoppard. She has coedited three volumes of the Study of Time series, including the most recent, *Time in Variance* (with Arkadiusz Misztal and Paul A. Harris, 2021).

Kathryn Prince is dean of the Faculty of Arts and associate professor of theater at the University of Ottawa and a former associate professor and visiting researcher at the Centre for the History of Emotions at the University of Western Australia. Her current funded research focuses on hopeful postapocalyptic narratives in various genres, building on her earlier work focused primarily on emotions in the theater. Her publications include *The Arden Research Handbook of Shakespeare in Contemporary Performance* (with Peter Kirwan, 2021), *Shakespeare in Canada: Remembrance of Ourselves* (with Irena R Makaryk, 2017), *History, Memory, Performance* (with David Dean and Yana Meerzon, 2014), and *Performing Early Modern Drama Today* (with Pascale Aebischer, 2012).

Sarah Jaquette Ray is chair and professor of environmental studies at California State Polytechnic University, Humboldt, and author of *The Ecological Other: Environmental Exclusion in American Culture* (2013) and *A Field Guide to Climate Anxiety: How to Keep Your Cool on a Warming Planet* (2020). Ray has coedited three volumes bridging the environmental humanities and social justice concerns and a fourth on pedagogy for uncertain times, *The Existential Toolkit for Climate Justice Educators.* She is a certified mindfulness teacher, offers workshops and speaks extensively on climate justice and emotions, and has written for the *Los Angeles Times, Scientific American, The Cairo Review of Global Affairs, Zócalo Public Square,* One Earth Sangha, and other venues.

Aaron Rosenberg is a lecturer in English literature at King's College London, where he has designed and taught an MA seminar called Unnatural Worlds: Literature and Ecology. He is the author of *Scale, Crisis, and the Modern Novel: Extreme Measures* (2023), and his recent scholarship has appeared in *NOVEL: A Forum on Fiction* and *Ecological Form: System and Aesthetics in the Age of Empire.*

Debra J. Rosenthal is professor of English at John Carroll University and author of *Performatively Speaking: Speech and Action in Antebellum American Literature* (2015) and *Race Mixture in Nineteenth-Century US and Spanish American Fictions* (2004). She is the editor of *Harriet Beecher Stowe's* Uncle Tom's Cabin: *A Routledge Sourcebook* (2003) and coeditor of *Cli-Fi and Class: Socioeconomic Justice in Contemporary American Climate Fiction* (2023), *Mixing Race, Mixing Culture: Inter-American Literary Dialogues* (2002), and *The Serpent in the Cup: Temperance in American Literature*

(1997). She is currently working on a manuscript called "Root Economy: Contemporary Climate Fiction and Class Politics." Rosenthal won John Carroll University's Culicchia Award for Excellence in Teaching and the Curtis W. Miles Award for Faculty Service.

Stephen Siperstein currently teaches at the Environmental Immersion Program at Choate Rosemary Hall. He also teaches courses in Choate's English department and multidisciplinary department and directs the school's writing center. In the summers, Siperstein codirects the Environmental Literature Institute at Phillips Exeter Academy. He is coeditor of *Teaching Climate Change in the Humanities* (2016), and his research focuses on developing effective strategies for interdisciplinary climate change education.

Ben Jamieson Stanley is assistant professor of English at the University of Delaware. Their scholarship focuses on how South African and Indian writers narrate relations between globalization and environmental precarity, often through the framework of food systems. Stanley's work is published in *ISLE: Interdisciplinary Studies in Literature and Environment, The Global South, The Oxford Handbook of Ecocriticism Online, Modernism and Food Studies* (2019), and *Cli-Fi and Class* (2023).

Orchid Tierney is assistant professor of English at Kenyon College. Her research focuses on waste, waste management, and atmospheric pollution in contemporary anglophone poetry. She is the author of *A Year of Misreading the Wildcats* (2019), and her scholarship, reviews, and poetry have appeared in *Jacket2, Venti, Fractured Ecologies*, and *SubStance*.

Rick Van Noy is professor of English at Radford University. He is the author of *Sudden Spring: Stories of Adaptation in a Climate-Changed South* (2018), shortlisted for the Association for the Study of Literature and the Environment's book award in creative writing, and *A Natural Sense of Wonder: Connecting Kids with Nature through the Seasons* (2018), honored with the Reed Environmental Writing Award by the Southern Environmental Law Center. *Borne by the River: Two Hundred Miles on the Delaware from Headwaters to Home* will be published in 2024.

Patrick Whitmarsh is visiting assistant professor of English at College of the Holy Cross and author of *Writing Our Extinction: Anthropocene Fiction and Vertical Science* (2023). His recent work has appeared or is forthcoming in *MFS: Modern Fiction Studies, Contemporary Literature*, and *SAF: Studies in American Fiction*.

Cynthia Schoolar Williams is associate professor in the School of Sciences and Humanities at Wentworth Institute of Technology in Boston and author of *Hospitality and the Transatlantic Imagination, 1815–1835* (2014). Her undergraduate elective Frozen! The Climate Crisis of 1816 and Its Lessons for Today won the 2018 Pedagogy Prize sponsored by the North American Society for the Study of Romanticism. From 2019 to 2021 Williams served as director of the Center for Sustainability and the Environment, a collaborative enterprise within the five-school consortium known as the Colleges of the Fenway.